Solar Energy Projects for the Evil Genius

Evil Genius Series

Solar Energy Projects for the Evil Genius

GAVIN D. J. HARPER

New York   Chicago   San Francisco   Lisbon   London   Madrid
Mexico City   Milan   New Delhi   San Juan   Seoul
Singapore   Sydney   Toronto

The **McGraw·Hill** Companies

Library of Congress Cataloging-in-Publication Data

Harper, Gavin D. J.

Solar energy—projects for the evil genius / Gavin Harper—1st

    p. cm — (Evil genius series)

    Includes index.

    ISBN 13: 978-0-07-147772-7

    ISBN 10: 0-07-147772-1 (alk. paper)

    1. Solar energy. I. Title.

TJ810. H3615 2007

621. 47078–dc22   2007020171

McGraw-Hill books are available at special quantity discounts to use as premiums and sales promotions, or for use in corporate training programs. For more information, please write to the Director of Special Sales, Professional Publishing, McGraw-Hill, Two Penn Plaza, New York, NY 10121-2298. Or contact your local bookstore.

3 4 5 6 7 8 9 0 DOC/DOC 0 1 2 1 0 9 8 7

ISBN 13: 978-0-07-147772-7

ISBN 10: 0-07-147772-1

This book is printed on acid-free paper.

Sponsoring Editor
Judy Bass

Editing Supervisor
David E. Fogarty

Project Manager
Andy Baxter

Proofreader
Grahame Jones

Indexer
Golden Paradox

Production Supervisor
Pamela A. Pelton

Composition
Keyword Group Ltd.

Art Director, Cover
Jeff Weeks

To the late Mr. P. Kaufman
who never failed to make science exciting

Gavin Harper is a sustainable technology advocate and popular author of how-to books. His other publications include *50 Awesome Auto Projects for the Evil Genius, Model Rocket Projects for the Evil Genius,* and *Build Your Own Car PC,* all for McGraw-Hill … and if you enjoyed the chapter on fuel cells, his forthcoming book *Fuel Cell Projects for the Evil Genius* will hit the shelves later this year. Gavin has had work published in the journal *Science* and has written for a number of magazines and online weblogs. His family continue to be bemused by his various creations, gadgets, and items of junk, which are steadily accumulating. He holds a BSc. (Hons) Technology with the Open University, and has completed an MSc. Architecture: Advanced Environmental & Energy Studies with UeL/CAT. He is currently studying towards a BEng. (Hons) Engineering with the Open University, and filling in spare time with some postgraduate study at the Centre for Renewable Energy Systems Technology at Loughborough University. He is rarely bored.

Gavin lives in Essex, United Kingdom.

Contents

Contents

Foreword

Gavin Harper's book *Solar Energy Projects for the Evil Genius* is a "must read" for every sentient human on this planet with a conscience, a belief in the bottom line, or a simple belief in the future of humanity.

At a time when such a book should be offered as suggested reading for the 19-year-old Gavin Harper, he's bucking the trend by actually being the author. Okay, so he's written a book on solar energy you say, big deal you say. You would be wrong. Not only is this Gavin's fourth book, it is nothing short of pure genius.

To be able to write about solar energy is one thing. But to possess the ability to put the knowledge of solar energy into layman's terms, while including examples of do-it-yourself projects which make the practical applications obvious, gives this boy genius the "street cred" (industry savvy) he so very much deserves.

This is a "how-to" book, which debunks the myth that "these things are decades away," and, without exception, should be in every classroom under the same sun.

So crack this book, turn on your solar light, and sit back for a ride into our "present"… as in "gift" from God.

Willie Nelson

Acknowledgments

There are always a lot of thank-yous to be said with any book, and this one is no exception. There are a lot of people that I would like to thank immensely for material, inspiration, ideas, and help—all of which have fed in to make this book what it is.

First of all, a tremendous thank-you to the staff and students of the MSc. Architecture: Advanced Environmental & Energy Studies course at the Centre for Alternative Technology, U.K. I never cease to be amazed by the enthusiasm, passion, and excitement members of the course exude.

I'd like to say a big thank-you to Dr. Greg P. Smestad, for his help and advice on photochemical cells. Dr. Smestad has taken leading-edge research, straight from the lab, and turned it into an accessible experiment that can be enjoyed by young scientists of all ages. I would also like to thank Alan Brown at the NASA Dryden Flight Research Center for the information he provided on solar flight for Chapter 15.

Also a big thank-you to Ben Robinson and the guys at Dulas Ltd. for their help in procuring images, and for setting a great example by showing how companies can be sustainable and ethical.

I'd also like to thank Hubert Stierhof for sharing his ideas about solar Stirling engines, and Jamil Shariff for his advice on Stirling engines and for continuing to be inspirational.

Thanks also to Tim Godwin and Oliver Sylvester-Bradley at SolarCentury, and to Andrew Harris at Schuco for sharing with me some of their solar installations.

An immense thank-you to Dave and Cheryl Hrynkiw and Rebecca Bouwseman at Solarbotics for sharing their insight on little solar-powered critters, and for providing the coupon in the back of the book so that you can enjoy some of their merchandise for a little less.

A massive thank-you to Kay Larson, Quinn Larson, Matt Flood, and Jason Burch at Fuelcellstore.com for helping me find my way with fuel cells, and for being inspirational and letting me experiment with their equipment. It would also be wrong not to mention H_2 the cat, who was terrific company throughout the process of learning about fuel cells.

Also, many thanks to Annie Nelson, and Bob and Kelly King of Pacific Biodiesel for providing me with some amazing opportunities to learn about biodiesel.

Thanks to Michael Welch at *Home Power* magazine, and also to Jaroslav Vanek, Mark "Moth" Green, and Steven Vanek, the designers of the fantastic solar ice-maker featured in Chapter 5. Their solar-powered ice-maker has already proven its immense worth in the developing world … and if you guys at home start building them at home and switching off your air-con and freezers, they stand to be a big hit in the developed world as well.

A big thank-you to my grandfather, who has seen the mess upstairs and manages to tolerate it, to my grandmother who hears about the mess upstairs and does not realize its magnitude, and to Ella who does a good job of keeping the mess within sensible limits—and knows when to keep quiet about it. Thanks are also long overdue to my dad, who is always immensely helpful in providing practical advice when it comes to how to build things, and to my mum who manages to keep life going when I have got my head in a laptop.

A huge thank-you to Judy Bass, my fantastic editor in New York who has been great throughout the trials and tribulations of bringing this book to print, and to the tremendous Andy Baxter (and the rest of his team at Keyword) who has managed to stay cool as a cucumber and provide constant reassurance throughout the editing process.

Chapter 1

Why Solar?

Our energy

In everyday life, we consume a tremendous amount of energy. Our lives are styled around consumption—consumption of natural resources and consumption of energy.

Figure 1-1 dramatically illustrates where all of this energy goes.

These figures are for a U.K. lifestyle, but we can take this as being representative for people who live in the "developed world."

The bulk of our energy consumption goes on space heating—58%—this is something that can easily be provided for with passive solar design.

Next is water heating, which requires 24% of the energy which we use—again, we will see in this book how we can easily heat water with solar energy.

So already we have seen that we can meet 82% of our energy needs with solar technologies!

The next 13% of our energy is used to provide electrical power for our lights and home. In Chapter 10 on solar photovoltaics, we will see how we can produce clean electricity from solar energy with no carbon emissions.

The remaining 5% is all used for cooking—again we will see in this book how easy it is to cook with the power of the sun!

So we have seen that all of our energy needs *can* be met with solar technologies.

Why solar?

The short answer to this question, albeit not the most compelling is "Why not solar?"

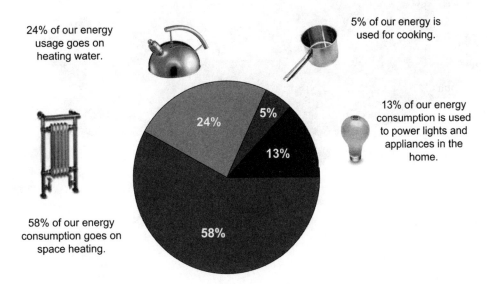

24% of our energy usage goes on heating water.

5% of our energy is used for cooking.

13% of our energy consumption is used to power lights and appliances in the home.

58% of our energy consumption goes on space heating.

Above is how UK household energy consumption can be split up into different uses.

Figure 1-1 *Domestic energy use. Information extracted from DTI publication "Energy Consumption in the United Kingdom." You can download this information from www.dti.gov.uk.*

Solar energy is clean, green, free, and best of all, isn't going to be going anywhere for about the next five billion years—now I don't know about you, but when the sun does eventually expire, I for one will be pushing up the daisies, not looking on with my eclipse glasses.

For the longer, more compelling answer, you are going to have to read the rest of this chapter. At the end, I hope that you will be a solar convert and be thinking of fantastic ways to utilize this amazing, environmentally friendly, Earth-friendly technology.

If we look at North America as an example, we can see that there is a real solar energy resource (Figure 1-2). While the majority of this is concentrated in the West, there is still enough solar energy to be economically exploited in the rest of the U.S.A.!

Renewable versus nonrenewable

At present, the bulk of our energy comes from fossil fuels—gas, coal, and oil. Fossil fuels are hydrocarbons, that is to say that if we look at them chemically, they are wholly composed of hydrogen and carbon atoms. The thing about hydrocarbons is that, when combined with the oxygen in the air and heat, they react exothermically (they give out heat). This heat is useful, and is used directly as a useful form of energy in itself, or is converted into other forms of energy like kinetic or electrical energy that can be used to "do some work," in other words, perform a useful function.

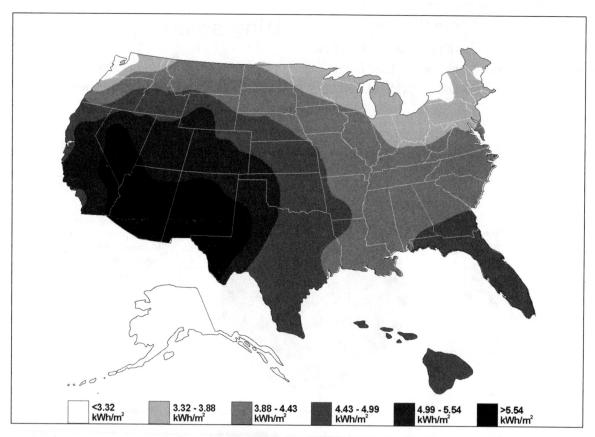

| <3.32 kWh/m² | 3.32 - 3.88 kWh/m² | 3.88 - 4.43 kWh/m² | 4.43 - 4.99 kWh/m² | 4.99 - 5.54 kWh/m² | >5.54 kWh/m² |

Figure 1-2   *North American solar resource.* Image courtesy Department of Energy.

So where did all these fossil fuels come from . . . and can't we get some more?

OK, first of all, the answer is in the question—fossils. Fossil fuels are so named because they are formed from the remains of animals and plants that were around a loooooong time ago. The formation of these fuels took place in the carboniferous period which in turn was part of the Paleozoic era, around 360 to 286 million years ago. This would have been an interesting time to live—the world was covered in lots and lots of greenery, big ferns, lush verdant forests of plants. The oceans and seas were full of algae—essentially lots of small green plants.

Although there are some coal deposits from when T-Rex was king, in the late cretaceous period around 65 million years ago, the bulk of fossil fuels were formed in the carboniferous period.

So what happened to make the fossil fuels?

Well, the plants died, and over time, layers of rock and sediment and more dead stuff built up on top of these carbon-rich deposits. Over many years, the tremendous heat and pressure built up by these layers compressed the dead matter

We have only recently started to worry about fossil fuels—surely we have time yet?

This is an incorrect assumption. For some time, people have prophesized the end of the fossil fuel age.

When the Industrial Revolution was in full-swing Augustin Mouchout wondered whether the supply of fossil fuels would be able to sustain the Industrial Revolution indefinitely.

"Eventually industry will no longer find in Europe the resources to satisfy its prodigious expansion. Coal will undoubtedly be used up. What will industry do then?"

Fossil fuel emissions

Take a peek at Figure 1-3. It is pretty shocking stuff! It shows how our fossil fuel emissions have increased dramatically over the past century—this massive amount of carbon dioxide in the atmosphere has dire implications for the delicate balance of our ecosystem and could eventually lead to runaway climate change.

Hubbert's peak and Peak Oil

Back in 1956 an American geophysicist by the name of Marion King Hubbert presented a paper to the American Petroleum Institute. He said that oil production in the U.S.A. would peak toward the end of the 1960s, and would peak worldwide in the year 2000. In fact, U.S. oil production did peak at the beginning of the 1970s, so this wasn't a bad prediction; however, the rest of the theory contains a dire warning.

The theory states that production of fossil fuels follows a bell-shaped curve, where production begins to gradually increase, then as the technology becomes mainstream there is a sharp upturn in production, followed by a flattening off when production has to continue against rising costs. As the costs of extraction increase, production begins to plateau, and then fall—falling sharply at first, and then rapidly.

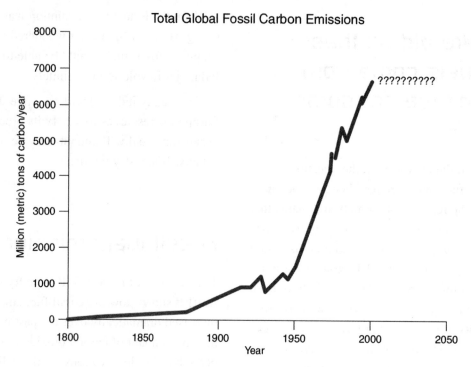

Total Global Fossil Carbon Emissions

Figure 1-3 *How our fossil fuel emissions have increased.*

This is illustrated in Figure 1-4.

This means that, if we have crossed the peak, our supplies of fossil fuels are going to begin to drop rapidly—when you think about how reliant we are on fossil fuels, this means that there is going to be a rapid impact on our way of life.

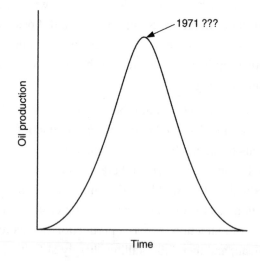

Figure 1-4 *Depiction of the "Peak Oil" scenario.*

So have we crossed the peak, and is there any evidence to support this?

The International Energy Agency has stated that energy production is in decline in 33 out of the 48 largest world oil producers. So, probably yes.

In the same way that there is Peak Oil, there is also Peak Coal, Peak Gas and Peak Uranium. All of these resources are in finite supply and will not last forever.

This means that those who believe that heavy investment in nuclear is the answer might be in for a shock. Nuclear has been touted by many as a means of plugging the "energy hole" left when fossil fuels run out; however, everyone in the world is facing the same problems—if everyone switches to nuclear power, the rate at which uranium is consumed will greatly increase.

A few other reasons why nuclear is a dumb option

Nuclear power really is pretty dangerous—talking about nuclear safety is a bit of a myth. Nuclear power stations are a potential target for terrorists, and if we want to encourage a clean, safe world, nuclear is not the way to go.

Nuclear makes *bad* financial sense. When the fledgling nuclear power industry began to build power stations, the industry was heavily subsidized as nuclear was a promising new technology that promised "electricity too cheap to meter." Unfortunately, those free watts never really materialized—I don't know about you, but my power company has never thrown in a few watts produced cheaply by nuclear power. Solar on the other hand is the gift that keeps on giving—stick some photovoltaics on your roof and they will pump out free watts for many years to come with virtually zero maintenance.

Decommissioning is another big issue—just because you don't know what to do with something when you finish with it isn't an argument to ignore it. Would you like a drum of nuclear waste sitting in your garden? All the world round, we haven't got a clue where to stick this stuff. The U.S.A. has bold plans to create Yucca mountain, a repository for nuclear waste—but even if this happens, the problem doesn't go away—it is simply consolidated.

Environmental responsibility

Until cheap accessible space travel becomes a reality, and let's face it, that's not happening soon, we only have one planet. Therefore, we need to make the most of it. The earth only has so many resources that can be exploited, when these run out we need to find alternatives, and where there are no alternatives then we will surely be very stuck.

Mitigating climate change

It is now widely acknowledged that climate change *is* happening, and that it *is* caused by man-made events. Of course, there is always the odd scientist, who wants to wave a flag, get some publicity and say that it is natural and that there is nothing we can do about it, but the *consensus* is that the extreme changes that we are seeing in recent times are a result of our actions over the past couple of hundred years.

Sir David King, the U.K.'s Chief Scientific Advisor says that climate change is "the most severe problem that we are facing today—more serious even than the threat of terrorism."

So how can we use solar energy?

When you start to think about it, it is surprising how many of the different types of energy sources around us actually come from the sun and solar-driven processes. Take a look at Figure 1-5 which illustrates this.

We can see how all of the energy sources in this figure actually come from the sun! Even the fossil fuels which we are burning at an unsustainable rate at the moment, actually originally came from the sun. Fossil fuels are the remains of dead animal and plant matter that have been subject to extreme temperature and pressure over millions of years. Those animals fed on the plants that were around at the time (and other animals) and those plants grew as a result of the solar energy that was falling on the earth.

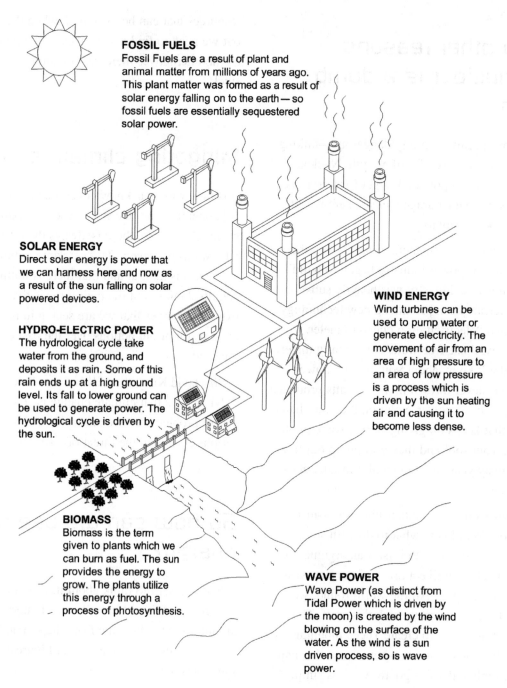

FOSSIL FUELS
Fossil Fuels are a result of plant and animal matter from millions of years ago. This plant matter was formed as a result of solar energy falling on to the earth—so fossil fuels are essentially sequestered solar power.

SOLAR ENERGY
Direct solar energy is power that we can harness here and now as a result of the sun falling on solar powered devices.

HYDRO-ELECTRIC POWER
The hydrological cycle take water from the ground, and deposits it as rain. Some of this rain ends up at a high ground level. Its fall to lower ground can be used to generate power. The hydrological cycle is driven by the sun.

WIND ENERGY
Wind turbines can be used to pump water or generate electricity. The movement of air from an area of high pressure to an area of low pressure is a process which is driven by the sun heating air and causing it to become less dense.

BIOMASS
Biomass is the term given to plants which we can burn as fuel. The sun provides the energy to grow. The plants utilize this energy through a process of photosynthesis.

WAVE POWER
Wave Power (as distinct from Tidal Power which is driven by the moon) is created by the wind blowing on the surface of the water. As the wind is a sun driven process, so is wave power.

Figure 1-5 *Energy sources.* Image courtesy Christopher Harper.

Biomass therefore is a result of solar energy—additionally, biomass takes carbon dioxide out of the atmosphere. When we burn it we simply put back the carbon dioxide that was taken out in the first place—the only carbon emissions are a result of processing and transportation.

Looking at hydropower, you might wonder how falling water is a result of the sun, but it is important to note that the hydrological cycle is driven by the sun. So we can say that hydropower is also the result of a solar-driven process.

Wind power might seem disconnected from solar energy; however, the wind is caused by air rushing from an area of high pressure to an area of low pressure—the changes in pressure are caused by the sun heating air, and so yet again we have another solar-driven process!

Tidal power is not a result of the sun—the tides that encircle the earth are a result of the gravitational pull that the moon has on the bodies of water that cover our planet. However, wave power which has a much shorter period, is a result of the wind blowing on the surface of the water—just as the wind is a solar-driven process, so is wave power.

So where does our energy come from at the moment?

Let's look at where the U.S.A. gets its energy from—as it is representative of many western countries.

If we look at the U.S.A.'s energy consumption, we can see (Figure 1-6) that most of our energy at the moment is produced from fossil fuels. This is a carbon-intensive economy which relies on imports of carbon-based fossil fuels from other countries, notably the Middle East. Unfortunately, this puts

America in a position where it is dependent on oil imported from other countries—politically, this is not the best position to be in. Next we look at hydro-power, which produces around 7% of America's electricity. Things like aluminum smelters, which require large inputs of electricity, are often located near to hydropower schemes because they produce an abundance of cheap electricity. Finally the "others" account for 5% of America's electricity production.

It is these "others" that include things such as solar power, wind powers and wave and tidal power. It is this sector that we need to grow in order to make energy supply more sustainable and decrease our reliance on fossil fuels.

This book is primarily concerned with development of the solar energy resource.

The nuclear lobby argue that nuclear is "carbon neutral" as the plants do not produce carbon dioxide in operation; however, this does not take into account the *massive* input of energy used to construct the plant, move the fuel, and decommission the plant. All of this energy (generally speaking) comes from high-carbon sources.

So we must look at the two remaining alternatives to provide our energy—hydro and "others."

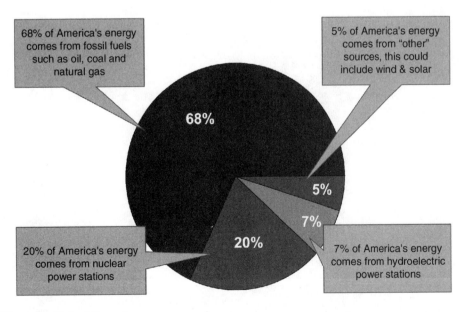

Figure 1-6 *Where the United States' energy comes from.*

There are limits to how much extra hydroelectric capacity can be built. Hydroelectricity relies on suitable geographic features like a valley or basin which can be flooded. Also, there are devastating effects for the ecosystems in the region where the hydro plant will be built, as a result of the large-scale flooding which must take place to provide the water for the scheme.

Micro-hydro offers an interesting alternative. Rather than flooding large areas, micro-hydro schemes can rely on small dams built on small rivers or streams, and do not entail the massive infrastructure that large hydro projects do. While they produce a lot less power, they are an interesting area to look at.

So all this is new right?

Nope . . . Augustin Mouchot, a name we will see a couple of times in this book said in 1879:

"One must not believe, despite the silence of modern writings, that the idea of using solar heat for mechanical operations is recent."

Why Solar?

8

The Solar Resource

The sun

Some 92.95×10^6 miles away from us, or for those working in metric 149.6×10^6 km away from us is the sun (Figure 2-1). To imagine the magnitude of this great distance, think that light, which travels at an amazing 299,792,458 meters per second, takes a total of 8.31 minutes to reach us. You might like to do a thought experiment at this point, and imagine yourself traveling in an airplane across America. At a speed of around 500 miles per hour, this would take you four hours. Now, if you were traveling at the speed of light, you could fly around the earth at the equator about seven and a half times in one second. Now imagine traveling at that speed for 8.31 minutes, and you quickly come to realize that it is a *long* way away.

Not only is it a long way away, but it's also pretty huge!

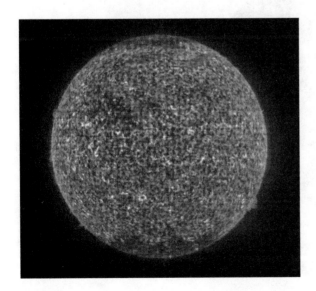

Figure 2-1 *The sun.* Image courtesy NASA.

It has a diameter of 864,950 miles; again, if you are working to metric standards that equates to 1.392 million km.

Although the sun is incredibly far away—it is also tremendously huge! This means that although you would think that relatively little solar energy reaches us, in fact, the amount of solar radiation that reaches us is equal to 10,000 times the annual global energy consumption. On average, 1,700 kWh per square meter is insolated every year.

Now doesn't it seem a silly idea digging miles beneath the earth's surface to extract black rock and messy black liquid to burn, when we have this amazing energy resource falling on the earth's surface?

As the solar energy travels on its journey to the earth, approximately 19% of the energy is absorbed by the atmosphere that surrounds the earth, and then another 35% is absorbed by clouds.

Once the solar energy hits the earth, the journey doesn't stop there as further losses are incurred in the technology that converts this solar energy to a useful form—a form that we can actually do some useful work with.

How does the sun work?

The sun is effectively a massive nuclear reactor. When you consider that we have such an incredibly huge nuclear reactor in the neighborhood already, it seems ridiculous that some folks want to build more!

The sun is constantly converting hydrogen to helium, minute by minute, second by second.

But what stops the sun from exploding in a massive thermonuclear explosion?—simple gravity! The sun is caught in a constant struggle between wanting to expand outwards as a result of the energy of all the complex reactions occurring inside it, and the massive amount of gravity as a result of its enormous amount of matter, which wants to pull everything together.

All of the atoms inside the sun are attracted to each other, this produces a massive compression which is trying to "squeeze" the sun inwards. Meanwhile, the energy generated by the nuclear reactions taking place is giving out heat and energy which wants to push everything outwards. Luckily for us, the two sets of forces balance out, so the sun stays constant!

Structure of the sun

Figure 2-2 illustrates the structure of the sun—now let's explain what some of those long words mean!

Starting from the center of the sun we have the core, the radiative zone, the convective zone, the photosphere, the chromosphere, and the corona.

The core

The core of the sun possesses two properties which create the right climate for nuclear fusion to occur—the first is incredibly high temperature 15 million degrees Celsius (I don't envy the poor chap who had to stand there with a thermometer to take the reading) and the second is incredibly high pressure. As a result of this nuclear fusion takes place.

In nuclear fusion, you take a handful of hydrogen nuclei—four in fact, smash them together and end up with one helium nucleus.

There are two products of this process—gamma rays which are high-energy photons and neutrinos, one of the least understood particles in the universe, which possess no charge and almost no mass.

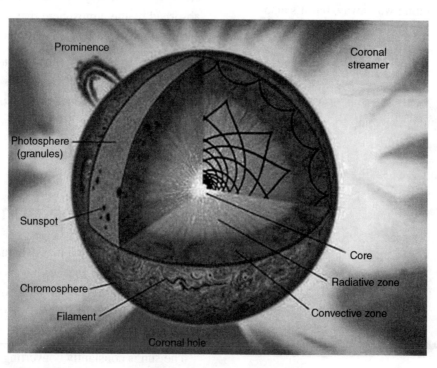

Figure 2-2 *The structure of the sun.* Image courtesy NASA.

The radiative zone

Next out from the core is the radiative zone. This zone is so named because it is the zone that emits radiation. A little bit cooler, the temperature in the radiative zone ranges from 15 million to 1 million degrees Celsius (even at that temperature though, I still wouldn't have liked to have been the one holding the thermometer).

What is particularly interesting about the radiative zone, is that it can take millions of years for a photon to pass through this zone to get to the next zone, aptly named the convective zone!

The convective zone

This zone is different, in that the photons now travel via a process of convection—if you remember high school physics, you will recollect that convection is a process whereby a body makes its way to a region of lower temperature and lower pressure. The boundary of this zone with the radiative zone is of the order of a million degrees Celsius; however, toward the outside, the temperature is only a mere 6,000°C (you still wouldn't want to hold the thermometer even with asbestos gloves).

The photosphere

The next region is called the photosphere. This is the bit that we see, because this is the bit that produces visible light. Its temperature is around 5,500°C which is still mighty hot. This layer, although relatively thin in sun terms is still around 300 miles thick.

The chromosphere

Sounding like a dodgy nightclub, the chromosphere is a few thousand miles thick, and the temperature rises in this region from 6,000°C to anywhere up to 50,000°C. This area is full of excited hydrogen atoms, which emit light toward the red wavelengths of the visible spectrum.

The corona

The corona, which stretches for millions of miles out into space, is the outer layer of the sun's atmosphere. The temperatures here get mighty hot, in fact up to a million degrees Celsius. Some of the features on the surface of the sun can be seen in Figure 2-2, but they are described in more detail in the next section and Figure 2-3.

Features of the sun

Now we have seen the inner machinations of the sun, we might like to take a look at what goes on on the surface of the sun, and also outside it in the immediate coronal region.

Coronal holes form where the sun's magnetic field lies. Solar flares, also known as solar prominences, are large ejections of coronal material into space. Magnetic loops suspend the material from these prominences in space. Polar plumes are

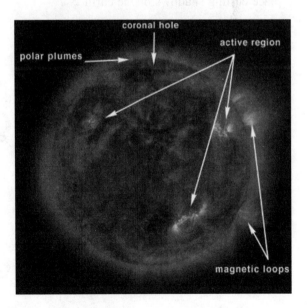

Figure 2-3 *Features on the sun's surface.* Image courtesy NASA.

altogether smaller, thinner streamers that emanate from the sun's surface.

The earth and the sun

Now we have seen what goes on at the source, we now need to explore what happens after that solar energy travels all the way through space to reach the earth's orbit.

Outside the earth's atmosphere, at any given point in space, the energy given off by the sun (insolation) is nearly constant. On earth, however, that situation changes as a result of:

- The earth changing position in space
- The earth rotating
- The earth's atmosphere (gases, clouds, and dust)

The gases in the atmosphere remain relatively stable. In recent years, with the amount of pollution in the air, we have noticed a phenomenon known as global dimming, where the particulate matter resulting from fossil fuels, prevents a small fraction of the sun's energy from reaching the earth.

Clouds are largely transient, and pass from place to place casting shadows on the earth.

When we think about the earth and its orbit, we can see how the earth rotates upon its axis, which is slightly inclined in relation to the sun. As the earth rotates at a constant speed, there will be certain points in the earth's orbit when the sun shines for longer on a certain part of the earth—and furthermore, because of the earth's position in space, that part of the world will tend to be nearer to the sun on average over the period of a day. This is why we get the seasons—this is illustrated in Figure 2-4.

As a result of the sun appearing to be in a different place in the sky, we may need to move our solar devices to take account of this. Figure 2-5 illustrates how a flat plate collector may need to be moved at different times of the year to take account of the change in the sun's position in order to harness energy effectively.

So how can we harness solar energy?

Thinking about it, more or less all of our energy has come either directly or indirectly from the sun at one point or another.

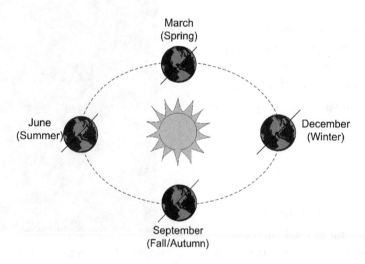

Figure 2-4 *The sun and seasons.*

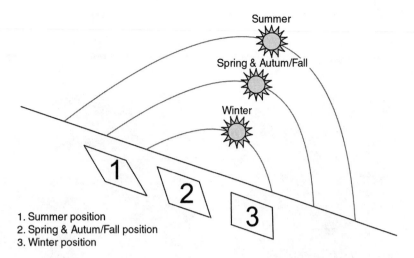

1. Summer position
2. Spring & Autum/Fall position
3. Winter position

Figure 2-5 *The sun changes position depending on the time of year.*

Solar power

Solar-powered devices are the most direct way of capturing the sun's energy, harnessing it, and turning it into something useful. These devices capture the sun's energy and directly transform it into a useful energy source.

Wind power

The heat from the sun creates convective currents in our atmosphere, which result in areas of high and low pressure, and gradients between them. The air rushing from place to place creates the wind, and using large windmills and turbines, we can collect this solar energy and turn it into something useful—electricity.

Hydropower

The sun drives the hydrological cycle, that is to say the evaporation of water into the sky, and precipitation down to earth again as rain. What this means is that water which was once at sea level can end up on higher ground! We can collect this water at a high place using a dam, and then by releasing the water downhill through turbines, we can release the water's gravitational potential energy and turn it into electricity.

Biomass

Rather than burning fossil fuels, there are certain crops that we can grow for energy which will replace our fossil fuels. Trees are biomass, they produce wood that can be burnt. Sugarcane can also be grown and be turned into bio-ethanol, which can be used in internal combustion engines instead of gasoline. Oils from vegetable plants can in many cases be used directly in diesel engines or reformed into biodiesel. The growth of all of these plants was initiated by the sun in the first place, and so it can be seen that they are derived from solar energy.

Wave power

Wave power is driven by the winds that blow over the surface of large bodies of water. We have seen how the wind is produced from solar energy; however, we must be careful to distinguish wave power from *tidal* power, which is a result of the gravitational attraction of the moon on a large body of water.

Figure 2-6   *Harnessing renewable energy to meet our energy needs cleanly.*

Figure 2-7   *Solar energy being harnessed directly on the roofs of the eco-cabins at the Centre for Alternative Technology, U.K.*

Fossil fuels

You probably never thought that you would hear an environmentalist saying that fossil fuels are a form of solar energy—well think again! Fossil fuels are in fact produced from the clean energy of the sun—at the end of the day, all they are is compressed plant matter which over millions of years has turned into oil, gas, and coal—and herein lies the problem. It took *millions* of years for these to form, and they are soon exhausted if we burn them at their present rate. So yes, they are a result of solar energy, but we must use them with care!

As we have seen, there are many ways in which we can harness solar power. Figure 2-6 shows some clean renewable ways in which we can capture solar energy not only from solar panels, but also from the power in the wind. Although not immediately apparent, the black pipeline that runs through the picture is in fact a small-scale hydro installation—yet another instance of solar energy being harnessed (indirectly).

This book focuses solely on "directly" capturing solar energy. In Figure 2-7 we can see a variety of technologies being used to capture solar energy directly in a domestic setting.

Chapter 3

Positioning Your Solar Devices

It is important to note that the position of the sun in the sky changes from hour to hour, day to day, and year by year. While this might be interesting, it is not very helpful to us as prospective solar energy users, as it presents us with a bit of a dilemma—where exactly do we point our solar device?

The ancients attributed the movement of the ball of fire in the sky to all sorts of phenomena, and various gods and deities. However, we now know that the movement of the sun through the sky is as a result of the orbital motion of the earth, not as a result of flaming chariots being driven through the sky on a daily basis!

In this chapter, we are going to get to grips with a couple of concepts—that the position of the sun changes relative to the time of the day, and also, that that position is further influenced by the time of the year.

How the position of the sun changes over the day

The ancients were aware of the fact that the sun's position changed depending on the time of the day. It has been speculated that ancient monuments such as Stonehenge were built to align with the position of the sun at certain times of the year.

The position of the sun is a reliable way to help us tell the time. The Egyptians knew this, the three Cleopatra's needles sited in London, Paris, and New York were originally from the Egyptian city of "Heliopolis" written in Greek as Ἡλίου πόλις. The name of the city effectively meant "town of the sun" and was the place of sun-worship.

It sounds like the destination for a pilgrimage for solar junkies worldwide!

We can be fairly sure that the obelisks that they erected, such as London's Cleopatra's needle (Figure 3-1), were used as some sort of device that indicated a time of day based on the position of the sun.

If you dig a stick into the ground, you will see that as the sun moves through the sky, so the shadow will change (Figure 3-2). In the morning the shadow will be long and thin; however, toward the middle of the day, the position of the shadow not only changes, but the shadow shortens. Then at the end of the day, the shadow again becomes long.

Of course, this effect is caused by the earth spinning on its axis, which causes the position of the sun in the sky to change relative to our position on the ground.

We will use this phenomena to great effect later in our "sun-powered clock."

How the position of the sun changes over the year

The next concept is a little harder to understand. The earth is slightly tilted on its axis; as the earth rotates about the sun on its 365¼-day cycle, different parts of the earth will be exposed to the sun for a longer or shorter period. This is why our days are short in the winter and long in the summer.

Figure 3-1 *Cleopatra's needle—an early solar clock?*

Figure 3-2 *How shadows change with the time of day.*

The season in the northern hemisphere will be exactly the opposite to that in the southern hemisphere at any one time.

We can see in Figure 3-3 that because of this tilt, at certain times of year, depending on your latitude you will receive more or less sunlight per day. Also if you look at your latitude relative to the sun, you can see that as the earth rotates your angle to the sun will be different at any given time of day, depending on the season.

We can see in Figure 3-4 an example house in the *southern* hemisphere—here we can see that the

sun shines from the north rather than the south . . . obviously if your house is in the northern hemisphere, the sun will be in the south!

This graphically demonstrates how the sun's path in the sky changes relative to your plot at different times of year, as well as illustrating how our rules for solar positioning are *radically* different depending on what hemisphere we are in.

What does this mean for us in practice? Essentially, it means that we need to change the position of our solar devices if we are to harness the most solar energy all year round.

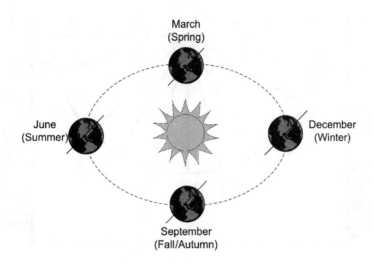

Figure 3-3 *How the earth's position affects the seasons.*

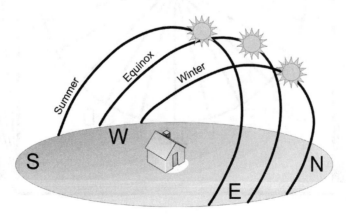

Figure 3-4 *Seasonal variation of the sun's position.*

Project 1: Build a Solar-Powered Clock!

You will need

- Photocopy of Figure 3-5
- Matchstick
- Glue

Tools

- Scissors

This is a dead-easy and quick sundial for you to build. Take a photocopy of Figure 3-5. If you want

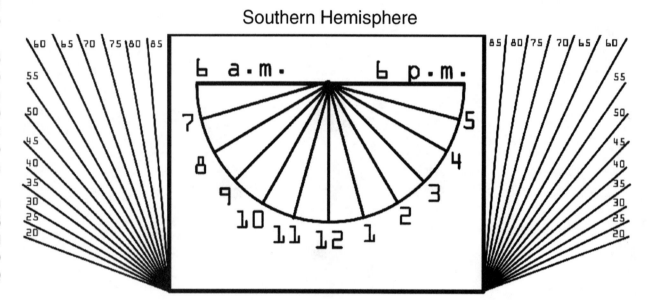

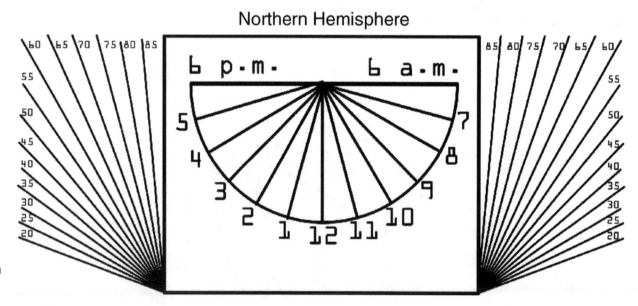

Figure 3-5 *Template for our "solar-powered clock."*

20

to be really flashy you can stick it to a piece of cardboard in order to make it more rigid and durable.

You need to cut out the dial that relates to the hemisphere that you are in—north or south. Then, you need to think about your latitude in degrees north or south. You will need to fold the sidepieces at the same angle marked in degrees as your latitude.

Stick a matchstick through the point at which all of the lines cross. What you should be left with is a piece of cardboard which makes an angle to the horizontal.

Now take your sundial outside and point the matchstick in the direction of due north (or south). You should be able to read the time off of the dial—compare this to the time on an accurate watch—remember you might have to add or take away an hour!

Rules for solar positioning

It is an artist's rule that you look more than you paint—for solar positioning this is also true. You need to look carefully and make observations in order to understand your site. Look at how objects on your plots cast shadows. See where your house overshadows and where it doesn't at various times of the year—remember seasonal variation—the position of the sun changes with the seasons and won't stay the same all year round (Figure 3-6).

Also, just because an area is shaded in one season, doesn't necessarily mean that it is shaded in all seasons. In fact, this can often be used to

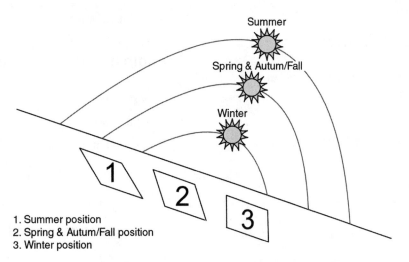

Summer

Spring & Autum/Fall

Winter

1. Summer position
2. Spring & Autum/Fall position
3. Winter position

Figure 3-6 *How seasonal variation affects the optimal position of solar collectors.*

your advantage. For example, in summer, you don't want too much solar gain in your house as it might overheat; however, in winter that extra solar energy might be advantageous!

Think carefully about trees—if they are deciduous, they will be covered with a heavy veil of leaves in the summer; however, they will be bare in the winter. Trees can be used a bit like your own automatic sunshade—in summer their covering of leaves blocks the sun; however, in the winter when they are bare they block less sun.

Make a record of your observations—drawings are great to refer back to. Keep a notebook where you can write any interesting information about what areas are and aren't in shadow. Note anything interesting, and the time of day and date.

Make sure that you are on the lookout on the longest and shortest days of the year—the first day of summer and the first day of winter. This is because they represent the extremes of what your solar observations will be; therefore, they are particularly useful to you!

Think about when in the day you will be using your solar device. Is it a photovoltaic cell that you would like to be using for charging batteries all day. Or, is it a solar cooker that you will be using in the afternoon? Think about when you want to use it, and what sunlight is available in what areas of your plot.

Work out which direction is north—try and find "true north" not just magnetic north. A compass will veer toward magnetic north so you need to find a way of compensating for this. Having a knowledge of where north and south is can be essential when positioning solar devices. Note which walls face which cardinal directions (compass points). If you are in the northern hemisphere, site elements where coolness is required to the north, and elements where heat is required to the south.

Think about the qualities of morning sun and evening sun. Position elements that require cool morning sun to the east—and those elements which require the hot afternoon sun to the west.

Project 2: Build Your Own Heliodon

You will need

For the cardboard heliodon

- Three rigid sheets of corrugated cardboard, 2 ft × 2 ft (60 cm × 60 cm)
- Packing tape
- Split leg paper fastener

For the wooden heliodon

- Three sheets of 1/2 in. (12 mm) MDF or plywood 2 ft × 2 ft (60 cm × 60 cm)
- Length of piano hinge 2 ft (60 cm)

- Countersunk screws to suit hinge
- Lazy Susan swivel bearing

For *both* heliodons, you will need

- Clip on spotlamp
- Length of dowel
- Large blob of plasticine/modeling clay

Tools

For the cardboard heliodon

- Scissors

- Craft knife
- Protractor

For the wooden heliodon

- Bandsaw
- Pillar drill
- Sander
- Protractor

We have already seen in this chapter about the sun's path—and we have learnt how we can use the sun to provide natural lighting and heating.

We saw in Figure 3-3 how the position of the sun and the earth influences the seasons, and how the path of the sun in the sky changes with the seasons. This is important to us if we want to design optimal solar configurations, as in order to maximize solar gain, we need to know where the sun is shining!

A heliodon is a device that allows us to look at the interaction of the light coming from the sun, and any point on the earth's surface. It allows us to easily model the angle at which the light from the sun will hit a building, and hence see the angle cast by shadows, and gauge the paths of light into the building.

The heliodon is a very useful tool to give us a quick reckoning as to the direction of light coming into the room, and what surfaces in that room will be illuminated at that time and date with that orientation.

A heliodon is also very useful for looking at overshadowing—seeing if objects will be "in the way" of the sun.

With our heliodon, it is possible to construct scale models that allow us to see, for example, if a certain tree will overshadow our solar panels. The heliodon is therefore a very useful tool for solar design, without having to perform calculations.

In this project, we present two separate designs. The first is for a cardboard heliodon, which is simple if you just wish to experiment a little with how the heliodon works. The design requires few materials and only a pair of scissors—but, it may

wear out over time. This does not mean that there is any reason for it to be less rigid than its sturdier wooden equivalent. The second design is for a more rigid permanent fixture which can be used professionally, for example if you are a professional who will routinely be performing architectural design or using the heliodon for education.

Our heliodon will consist of three pieces of board. The first forms a base; on top of this base, we affix a second board which is allowed to swivel by way of, in the wooden version, a "Lazy Susan" bearing. This is a ball-bearing race that you can buy from a hardware shop, which is ordinarily used as a table for a "Lazy Susan" rotating tray.

In the cardboard version, we simply use a split leg pin pushed through the center of both sheets, with the legs splayed and taped down.

The third board is hinged so that the angle it makes with the horizontal can be controlled, it is also equipped with a stay to allow it to be set at the angle permanently and rigidly. And that is just about it! With the wooden version, a length of piano hinge accomplishes this job admirably, and with the cardboard version, a simple hinge can be made using some strong tape.

The other part of the heliodon is an adjustable light source. This can be made in a number of ways. The simplest of which is a small spotlamp equipped with a clip that allows it to be clamped to a vertical object such as the edge of a door. Slide projectors are very good at providing a parallel light source—these present another option if their height can easily be adjusted. If you will be using the heliodon a lot, it would make sense to get a length of wood mounted vertically to a base, with the dimensions given in Table 3-1 marked permanently on the wood.

Heliodon experiments

Once you have constructed your heliodon you can begin to perform some experiments using it.

Table 3-1

Lamp heights for different months of the year

January 21	8 in.	20 cm	from floor
February 21	22 in.	55 cm	from floor
March 21	40 in.	100 cm	from floor
April 21	58 in.	145 cm	from floor
May 21	72 in.	195 cm	from floor
June 21	80 in.	200 cm	from floor
July 21	72 in.	195 cm	from floor
August 21	58 in.	145 cm	from floor
September 21	40 in.	100 cm	from floor
October 21	22 in.	55 cm	from floor
November 21	8 in.	20 cm	from floor
December 21	2 in.	5 cm	from floor

These measurements are assuming a measurement of 87 in. between the center of the heliodon table and the light source

You need to be aware of the three main adjustments that can be made on the heliodon.

- Seasonal adjustment—by moving the lamp up and down using the measurements listed above, it is possible to simulate the time of year.

- Latitude adjustment—by setting the angle that the uppermost flat sheet makes with the base, you can adjust the heliodon for the latitude of your site.

- Time of day adjustment—by rotating the assembly, you can simulate the earth's rotation on its axis, and simulate different times of day.

The two table adjustments are illustrated in Figure 3-7.

In order to secure the table at an angle, probably the easiest way is to use a length of dowel rod with a couple of big lumps of modeling clay at each end. Set the angle of the table to the horizontal, then use the dowel as a prop with the plasticine to secure and prevent movement.

There are a couple of simple experiments that we can do with our heliodon to get you started. Remember the sundial that you made earlier in the book? Well, set the angle of latitude on your table to the angle that you constructed your sundial for (Figure 3-8). You will see that as you rotate the table, the time on the sundial changes. You can use

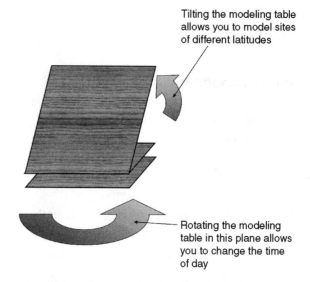

Figure 3-7 *Heliodon table adjustments.*

Tilting the modeling table allows you to model sites of different latitudes

Rotating the modeling table in this plane allows you to change the time of day

Compass points

Remember to think carefully about where north and south are in relation to your modeling table. Consider whether the site you are modeling is in the north or south hemisphere and adjust the position of your model accordingly.

this approach to calibrate your heliodon. You might like to make some marks on the cardboard surface to indicate different times of day.

The next stage of experimentation with the heliodon is to look at modeling a real building.

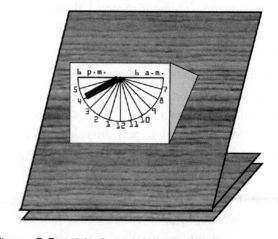

Figure 3-8 *Heliodon sundial experiment.*

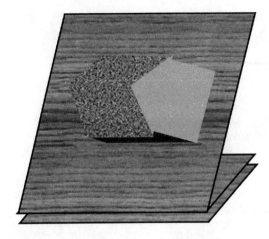

Figure 3-9 *Using a cardboard model building to model solar shading.*

Construct a model from cardboard (Figure 3-9), and include for example, window openings, doors, patio doors, and skylights. By turning the table through a revolution, it is possible to see where the sun is penetrating the building, and what parts of the room it is shining on. This is useful, as it allows us to position elements of thermal mass in the positions where they will receive the most solar radiation.

We can also make models of say, a solar array, and cluster of trees, and see how the trees might overshadow the solar array at certain times during the year. Use the heliodon with scale models to devise your own solar experiments!

Now with modern computer aided design (CAD) technology, the heliodon can be replicated digitally inside a computer. Architects routinely use pieces of CAD software to look at how light will penetrate their buildings, or whether obstructions will overshadow their solar collectors. However, heliodons are still a very quick, simple technology which can be used to make a quick appraisal of solar factors on a model building. A professional, more durable heliodon can be seen in Figure 3-10.

Project 3: Experimenting with Light Rays and Power

You will need

- Small torch
- Length of string
- Tape
- Big sheet of paper
- Bunch of pencils
- Elastic band

Attach the large sheet of paper to the wall using the tape. Then, take the piece of string, and attach one end roughly to the center of the paper with the tape. Now hold the string to one side of the piece of paper, and attach the torch to the string so that the bulb of the torch falls within the boundary of the paper.

We are going to see how angle affects the light power falling on a surface when the distance from the surface remains the same.

Now imagine our torch as the sun, hold the torch to face the paper directly keeping the string taught. You should see a "spot" of light on the paper.

Figure 3-10 *A professional architect using a heliodon to make estimations of solar gain on a model building.*

Draw a ring around the area of highest light intensity. Now, hold the torch at an angle to the paper, and again with the string taught, draw a ring around the area of high intensity. Repeat this at both sides of center a few times at different angles.

Figure 3-11 shows us what your sheet of paper might look like.

What can we learn from this? Well, the power of our torch remained the same, the bulb and batteries were the same throughout the experiment, the amount of light coming out of the torch did not change.

However, the area on which the light fell did change. When the torch was held perpendicular to the paper, there was a circle in the middle of the page. However, hold the torch at an angle to the page and the circle turns into an oval—with the result that the area increases. What does this mean to us as budding solar energy scientists? Well, the sun gives out a fixed amount of light; however, as it moves through the sky, the plane of our solar collectors changes in relation to the position of the sun. When the sun is directly overhead of a flat plate, the plate receives maximum energy; however, as we tilt the plate away from facing the sun directly, the solar energy reaching the plate decreases.

You might have noticed that as you angled the torch and the beam spread out more, the beam also became dimmer.

Remember the bunch of pencils? Well grab them and put an elastic band around them. Imagine each pencil is a ray of light from the sun. Point them down and make a mark with the leads on a piece of paper. Now, carefully tilt all the pencils in relation to the paper and make another mark with all the pencils at the same time (Figure 3-12). As you can see, the marks are more spread out. Remembering that we are equating our pencil marks with "solar rays," we can see that when a given beam of light hits a flat surface, if the beam hits at an oblique angle, the "rays" are more spread out. This means that the power of the beam is being spread out over a larger area.

It is important that we understand how to make the most of the solar resource in order to make our solar devices as efficient as possible.

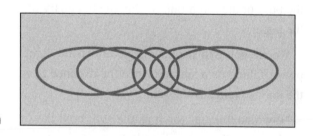

Figure 3-11 *Light ray patterns drawn on paper.*

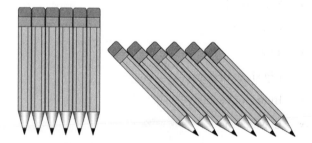

Figure 3-12 *Bunch of pencils experiment.*

Chapter 4

Solar Heating

The sun provides us with heat and light that is essential to life all year round.

One of the most efficient ways of harnessing the sun's energy is to use it to space heat our buildings, and produce hot water for our daily needs, such as washing, cleaning, and cooking.

When you think about the truly tremendous amount of heat that the sun produces, it seems absolutely ridiculous that we should want to burn our precious fossil fuels to heat things up.

We can use the sun to directly heat our buildings—this is known as passive heating—or we can use an intermediate storage and distribution medium such as water or air. The advantage of using water or air as a storage medium for the heat, is that we can concentrate the sun, and collect it efficiently using solar collectors, and then using a distribution network of pipes or ducts, we can direct the heat to where we want it; and, more importantly, direct the heat to the places where it can be utilized most effectively.

In this chapter, we are going to be looking at the fundamentals of a solar hot water heating system. By the end of the chapter, you should have an understanding of how such systems work, and be armed with the knowledge to begin researching and installing your own hot water system.

Why use solar energy for heating?

There are considerable environmental benefits associated with using renewable energy for heating.

Consumption of fossil fuels for heating is tremendous when you consider the global scale. Producing as much as possible of our heat from renewable resources will considerably reduce our consumption of fossil fuels.

Can I use my roof to mount my solar heating panels?

The roof seems an obvious place to want to mount your solar heating panels. After all, you have a large area which is currently unutilized just waiting for some clean green energy generation!

First of all, you should consider the structural integrity of your roof and how strong it is. Remember, the roof will not only need to support the weight of the solar heating panel and all of the associated paraphernalia, but might also need to support your weight as you install it.

You will also need to consider the orientation of your roof and whether it is positioned in such a manner that it will receive optimal solar gain. If you are in the northern hemisphere, you will want a roof which faces as near to due south as possible. If your roof does not face directly due south, there will be some loss of efficiency—which is proportional to the angle of deviation from due south.

If you live in the southern hemisphere, the reverse is true—you want a roof that faces due north in order to catch the best of the sun's rays.

How does solar heating work?

On a hot summer day, if you are walking around a parking lot, gently touch a black car, and the chances are it will feel *very* hot. Now touch a silver or white car, and you will find that it is significantly cooler.

This is the principle that underpins solar heating. A black surface heats up quickly in the sun.

Our demand for hot water is driven by a number of things. We use hot water every day for tasks such as washing our hands, clothes and dishes. From now on, we will refer to this as "solar hot water." We can also use hot water for heating our homes. We will refer to this as "solar space heating" from now on.

What we need to do, is look at our demand for heated water, and see how it correlates to the energy available from the sun.

Solar hot water

Our demand for hot water is fairly constant throughout the year. We use more or less the same amount of hot water for washing and cleaning in the winter as we do in the summer.

Solar space heating

We can also use solar energy to heat our space directly—passively, rather than using an active system. This is called passive solar design. We can design our buildings with large expanses of glass on the sun-facing façades in order to capture the solar energy and keep the building warm and light. However, the requirements for space heating are different in the winter from in the summer. If we design our buildings for "summer conditions," they could be intolerably cold in the winter. For this reason, we can use architectural devices such as shading and brie soleil to ensure that the room receives an optimal amount of light in both summer and winter. Passive solar design is a whole book in its own right though!

What does a solar heating system look like?

Figure 4-1 illustrates a basic solar water heating system.

We can see a large storage tank in the Figure. This is filled with water and is used as a thermal store. It is imperative that this tank is incredibly well insulated as it is pointless going to a lot of effort to collect this solar energy if we then lose it in storage!

You will notice that the solar hot water tank has a gradient fill—this denotes the stratification of the water—the colder water sinks to the bottom, while the warmer water is at the top of the tank.

We draw the hot water off from the top of the tank, while replacing the hot water with cold water at the bottom of the tank. This allows us to maintain the "layered" stratified nature of the tank.

At the bottom of the tank, we can see a coil; this is shown more clearly in Figure 4-2—this coil is in fact a copper pipe—we can see that the pipe enters the tank at the bottom, and exits the tank at the top. The pipes are connected in a closed circuit to a solar collector. This closed circuit is filled with a fluid which transfers the heat from the solar cell to the tank.

This is the simplest type of solar system—it is called a thermosiphon. The reason for this name is that the process of circulation from the solar cell to the tank is driven by nothing more than heat. Natural convective currents set up a flow, whereby the hot water makes its way around the circuit.

It is also possible to insert a pump into this circuit to increase the flow of the heat transfer medium.

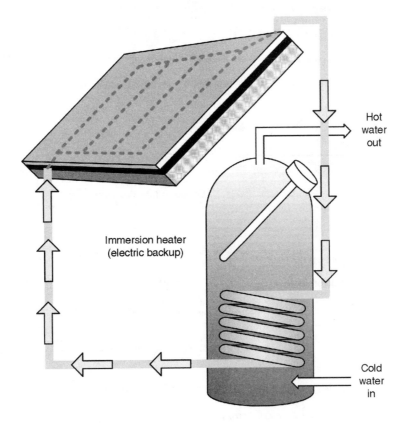

Hot
water
out

Immersion heater
(electric backup)

Cold
water
in

Figure 4-1 *A basic solar water heating system.*

We can also drive this pump using photovoltaic solar cells. This means that our heating is not using electricity from the grid—and hence not using energy generated from fossil fuel sources. There is one manufacturer, Solartwin, which supplies a system which consists of a solar thermal panel, and a pump driven by photovoltaics. The advantage of this approach is that the energy for the pump is provided at the same time as there is heat in the system.

Figure 4-2 *A cutaway of a thermal store tank.*

Tip

A good science fair project might be to build a demonstration solar water heating system using easy-to-use flexible aquarium tube for the "plumbing" and a soda bottle for the hot water storage tank. A few thermocouples or thermistors will allow you to monitor the temperatures around the setup and see how effectively it is working.

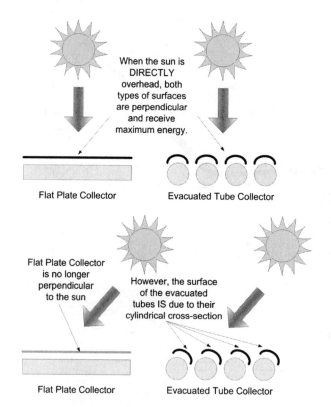

When the sun is DIRECTLY overhead, both types of surfaces are perpendicular and receive maximum energy.

Flat Plate Collector

Evacuated Tube Collector

Flat Plate Collector is no longer perpendicular to the sun

However, the surface of the evacuated tubes IS due to their cylindrical cross-section

Flat Plate Collector

Evacuated Tube Collector

Figure 4-3 *Flat plate versus evacuated tube collectors.*

Solar collectors

There are two types of solar collector: flat plate, and evacuated tube. We can see in Figure 4-3 the two types of collectors compared. While a greater amount of sun falls on the flat plate, the evacuated tube collectors are better insulated. However, as the sun moves in an arc through the sky, the flat plate collector's effective area becomes smaller, and as the evacuated tube collectors are cylindrical, the area presented toward the sun is the same.

In Figure 4-4 we see the make up of a flat plate collector. It is essentially quite a simple device. There is insulation, which stops the heat that it absorbs from being transmitted into the roof it is mounted on. A coil of tube within this collects the heat and transmits it to the storage tank, and at the front of the collector is an absorbent surface.

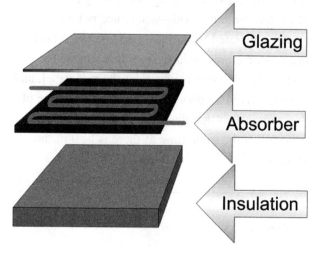

Figure 4-4 *Cutaway of a flat plate collector.*

This could simply be matt black, or it could be a selective coating.

On the roof shown in Figure 4-5 we can see a variety of different solar cells, both thermal and photovoltaic nestling together in harmony.

Figure 4-5   *An array of different solar thermal cells on a roof.*

Project 4: Build Your Own Flat Plate Collector

We are now going to make a flat plate collector. There are a number of different types of collector, all suitable for relatively simple manufacture in a home workshop (Figures 4-6 to 4-8).

The key thing to remember about solar collectors is keeping the heat in and the cold out. This can be accomplished by using glazing on the sun-facing side of the panels and thermal insulation on the side

Figure 4-6   *A commercially made clip fin collector.*

Figure 4-7 *A home-made clip fin collector.*

that faces away from the sun. We need to try to eliminate thermal bridges as far as we possibly can.

Aluminum clip fins are one of the easiest ways of assembling a solar collector quickly, as they essentially clip onto a matrix of copper pipe.

Another way of constructing a solar collector is to use an old radiator painted black inside an insulated box—crude but effective! (Figure 4-9). This system contains more water, and as a result has a slower response time. This is because it takes more time to heat up the thermal mass of the radiator.

Warning

One of the problems that solar collectors suffer from is freezing in the winter. When temperatures drop too low, the water in the pipes of the collectors expands—this runs the risk of *severely* damaging the collectors.

Figure 4-8 *Aluminum clip fins.*

Figure 4-9 *A recycled radiator collector.*

Project 5: Solar Heat Your Swimming Pool

While having a pool in your yard is a great way to exercise and enjoy the summer sun, swimming pools are notorious for "drinking" energy. The problem is that there is simply such a great volume of water to heat!

Energy is becoming more expensive as we begin to realize the serious limitations of the previously cheap and abundant fossil fuels.

Some people heat their pools in order to be able to enjoy them out of season; however, that comes with a big associated energy cost.

Before you even start to consider heating your pool using solar energy, you need to consider energy reduction and efficiency measures. You might want to consider your usage patterns. Will it really make much difference to me if I can't use my pool out of season? After all, who really wants to swim when it is cold and wet outside! Also you might want to consider energy minimization strategies. Is your pool outside and uncovered at the moment? Heat rises . . . so all that heat that you

are throwing into your pool is being lost as it dissipates into the atmosphere. This isn't smart! Building some kind of enclosure over your pool will make the most of any investment that you put into solar heating your pool.

Once you have taken steps to minimize the energy that your pool requires, you can begin to make advances toward heating it using free solar energy. There is nothing really too complicated about a solar pool heating system. As we only need to elevate the temperature of the water slightly, we can use simple unglazed reflectors.

The reason? Well think of it like this . . . the water you get from the hot tap to wash with is significantly hotter than the sort of temperature you would be expecting to swim in. A domestic solar hot water rig heats a small volume of water to a very high temperature. By contrast, a solar pool heating system, takes a large quantity of water, and heats it by a small amount. Here is the fundamental difference. Because the water is circulating at a faster rate, unglazed collectors can provide acceptable efficiency.

But that's not all!

In some hot climates, pools can have a tendency to overheat. Solar collectors can save the day here! By pumping water through the collectors at night it is possible to dump excess heat.

This technology isn't just applicable to small pools at home, large municipal pools are also heated by solar technology in a number of cases. Take for instance the International Swim Center at Santa Clara, California, 13,000 square meters of solar collector heat a total of 1.2 million gallons of water a day!

Figure 4-10 illustrates solar pool heating.

> ## Tip
>
> Enerpool is a free program that can be used to simulate your swimming pool being heated with solar collectors. By inputting information such as your location, and how the pool is covered. The program can predict what temperature your pool will be at, at any given time!
>
> www.powermat.com/enerpool.html

The Supplier's Index (Appendix B) lists a number of companies that sell products for solar heating your pool.

Do we need to use solar thermal power directly?

If we consider power generation on a large scale, all of our power stations whether they be nuclear, coal, oil, or gas fired, all produce heat primarily, and then use this heat to produce steam, which then, through using rotating turbines, produces electricity.

This means, that at present, we do not produce electricity directly from chemicals, like we do in a battery—we first produce heat as an intermediate process, which is in turn used to produce electricity.

Once we recognize this, we quickly realize that it could be possible to use solar thermal energy to raise steam to generate electricity.

And this is exactly what they are doing in Kramer Junction, California.

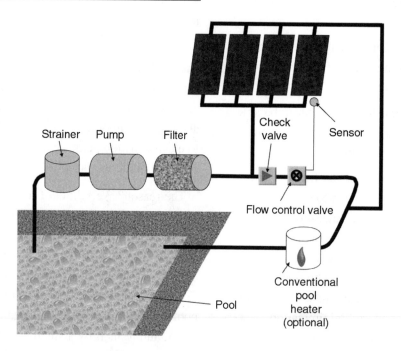

Figure 4-10   *Solar pool heating.*

Figure 4-11 *Solar thermal power in the Mojave desert.* Courtesy Department of Energy.

Project 6: Useful Circuits for Solar Heating

Although we have discussed basic systems here for producing solar heat—and by no means is this comprehensive coverage of solar heating (the subject really deserves several books in its own right) there are a number of things that we can do to improve our system. If our system is "active," which is to say if there is a pump driving a working fluid around the system, then we can do a little more to control the fluid.

If our system is passive, i.e., a thermosiphoning system, where the fluid makes its way around the system as a result of changes in density, then we might at least like some feedback as to what our system is doing.

You will need

- Negative temperature coefficient thermistor
- 2 × 10 k resistor
- 100 k variable resistor
- 741 op amp
- 1 M resistor
- 4.7 k resistor
- BC109 NPN transistor
- 6 V piezo buzzer
- Heatshrink tubing
- Mastic/silicone sealant

Optional

- 6 V relay
- Protection diode

Tools

- Soldering iron
- Side cutters
- Solder

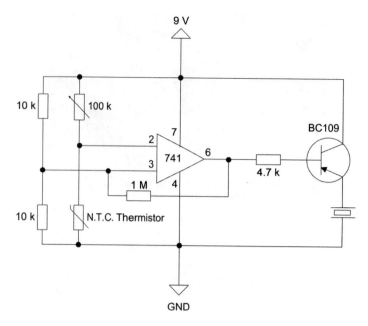

Figure 4-12 *Solar heating over temperature indicator.*

The following is a simple circuit that uses a thermistor as the sensing element to provide feedback as to the condition of a surface being monitored. There are two simple circuits here—both are similar with one small change—the position of the thermistor and variable resistor which change places (Figures 4-12 and 4-13).

Protecting the sensor against the elements

A thermistor as supplied from the components shop is a pretty fragile beast, and as such should be respected if reliable operation is wanted.

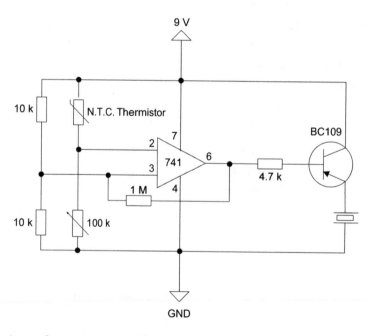

Figure 4-13 *Solar heating under temperature indicator.*

The device is designed to be soldered onto a printed circuit board; however, we are expecting it to be used in a much more hostile environment. For this reason, provision should be made to insulate the leads of the thermistor with some heatshrink tubing, which will provide mechanical support for the soldered joint, and also prevent ingress of water.

Once each individual lead has been insulated, the pair can be bound together with a little more heatshrink, or failing that a little bit of tape. The sensor should be provided with long enough leads to comfortably reach the circuit board. If your solar collector tracks the sun, you will need to ensure that the leads are sufficiently long to reach back to the circuit board, even at the extent of your collector's movement.

Taking this a step further, you need to mate your sensor to the surface being monitored. A little squeeze of heat transfer compound in the gap between thermistor and the surface is not a bad idea—you can get this from computer suppliers, as it is commonly used to ensure a good surface interaction between a CPU and heatsink.

Once you have done this, you can apply silicone sealant liberally to hold the sensor in place. If you really want to go the whole hog, you can even insulate the other side of the sensor with a little thermal lagging, such as polystyrene or a slice of foam pipe lagging. This will prevent the sensor from being unduly influenced by external fluctuations in temperature.

Calibrating the sensor

When setting the circuit up, you will need to calibrate the sensor against a reference of known temperature. A water bath is a good way of providing a stable temperature. To get the temperature you desire, get a cup of ice water and a cup of boiling water, and use these to adjust the temperature of a cup of water with a thermometer immersed in it.

Modifications to the circuit

Although the circuit is useful in its own right, there are a couple of things that we can do to improve its functionality. As it stands, the circuit provides an "alert" when the temperature goes out of condition; however, we might consider a scenario where we are not at home to take action. In this instance, we might want to replace the piezo buzzer in the system with a relay and protection diode. This is a straight swap, and allows the circuit to then control an automatic device which can take action—for example a pump or electronic valve.

To give you an example of how this circuit could be useful in your solar water heating system, in freezing weather, if no water is circulating in the pipes, there is a risk that that water could expand and burst your pipes. To overcome this, a relay could switch on a trickle pump which keeps a little water circulating through the pipes. This water will carry with it some heat from the thermal store, which should keep your pipes free of ice.

Equally, with the resistors reversed as in the second circuit, you might want to set up a system whereby a pump is triggered when heat in the collector is sensed. This ensures that hot water isn't pumped through a cold collector.

What is the future of solar heating?

It is inevitable that in the future, we will need to seek different solutions to our problems as we are forced not to depend on fossil fuels. Solar energy certainly has a place in meeting our heating needs in the future, and considering the energy from the sun is free, it is very surprising that more people aren't using it now!

We have seen in this chapter how solar heating can meet our heating needs, but that the availability of solar energy is seasonal, and in part that determines the supply of solar hot water.

Even when the sun cannot provide all of the energy all of the time, or where 100% solar provision would be uneconomical, it can certainly go a long way toward reducing the amount of energy we need to consume. Even pre-heating water a little bit in winter goes some way to saving energy.

Another thing that needs careful consideration, is that if we need to provide extra energy for our heating needs, where will that energy come from? Fossil fuels pollute the atmosphere and are a finite resource, nuclear leaves a legacy of toxic waste, but maybe solar energy has another trick up its sleeve—biomass!

The trees and plants that we are surrounded by are effectively solar batteries! They take the energy produced from the sun, and using a process called photosynthesis, use the energy to grow. All the time that they are growing, they are taking in carbon dioxide from the atmosphere and producing oxygen. Once a tree has grown, we can cut it down and burn it. While this process releases carbon dioxide into the atmosphere, there is no "net gain" of carbon dioxide, as the tree took the carbon dioxide out of the atmosphere while growing!

Chapter 5

Solar Cooling

In hot climates, it can often become uncomfortably hot—in the modern world, we tend to look toward air conditioning to provide a comfortable internal atmosphere; however, air conditioning often leaves us with dry stale air.

While it would seem to be counterintuitive to use the sun to cool things down, there are a number of techniques that we can use to cool things down by employing the sun's energy.

Why air conditioning is bad

The amount of energy that air conditioning consumes is truly tremendous. In addition to this, the heat extracted from the building is simply dumped out into the atmosphere. Air conditioning cooling stacks are a breeding ground for Legionella bacteria, and the refrigerants used in air conditioning are ozone-depleting and add to the burden of global warming. While there has been a worldwide move to eliminate CFCs from air conditioning units because of the damage they do, the interim HCFC and HFC chemicals are still not environmentally friendly.

What can we do instead?

Rather than using large amounts of fossil fuels, there are a number of other strategies that we can use to cool our buildings.

Passive solar cooling

There are a number of ways that we can design our buildings to stay at a pleasant internal temperature, and prevent them from overheating, even in the summer.

Trombe walls

As with many of the themes in this book, this idea is not a new one, in fact it was patented in 1881 (U.S. Patent 246626). However, the idea never really gained much of a following until 1964, when the engineer Felix Trombe and architect Jacques Michel began to adopt the idea in their buildings. As such, this type of design is largely referred to as a "Trombe wall."

Figure 5-1 shows a Trombe wall on one of the resident's houses at the Centre for Alternative Technology (CAT), U.K.

Let's describe the construction and operation of the Trombe wall.

Essentially, the Trombe wall is a wall with a high thermal mass, the wall is painted black to enable it to absorb solar radiation effectively. The wall is also separated from the outer skin of glazing by an air gap.

The original Trombe walls were not particularly effective. They worked by absorbing heat in the thermal mass during the day. At night, this heat would be dissipated both into the room—but also to the outside through the air gap and glazing. The theory was that the glazing would help to retain heat, and because the thermal mass had gained enough heat during the day, it would be warmer

Figure 5-1   *Trombe wall at CAT, U.K.*

will know that as air heats up, the molecules of gas gain a little more energy—this causes them to bounce around a bit more, and as a result, they tend to spread out a little bit. As they do this, the body of gas becomes less dense. As you will know if you have ever observed a spill of oil on a body of water, the less-dense compound floats to the surface as it is displaced by the more-dense compound. In this case, the lighter air rises up through the gap between the wall and the glazing. This is the principle that hot air balloons use to operate—hot less-dense air floats above denser air!

This convective current can be used to either heat or cool the building.

Remember those gaps in the wall and glazing—well, if both of the vents in the wall (the thermal mass side of things) are opened, air will be sucked out of the room at the bottom, heated as it contacts the thermal mass, and using convection will rise up to the top of the air gap, where it flows back into the room.

Of course during the summer, this heat isn't really wanted—so the flaps can simply be closed in order to keep the room cool.

But this chapter did say it was about solar cooling!

Well, if you also have flaps in the glazing which can be opened and closed, you can then open a flap at the top of the glazing, and at the bottom of the thermal mass. The flap at the top of the thermal mass is closed.

This sets up a convective current which sucks air out of the bottom of the room, and heats it slightly, causing it to rise. But rather than this air being fed back into the room, the air is instead dissipated into the outside atmosphere. This has the effect of sucking air from the room. This air has to be replaced somehow, so what happens is that fresh air is sucked in through cracks in the building fabric, gaps in doors and windows, etc. This provides a fresh cooling breeze for the occupants (Figure 5-2).

than the internal room temperature, as a result, the room would warm up.

In reality it appeared that most of the heat was simply dumped to the cold outside.

A series of improvements were made to the design of the Trombe wall which significantly increased its performance. In the improved version of the Trombe wall, there are vents at the top and bottom of the wall, and also on the glazing.

These vents have a mechanism that allows them to be opened and closed in certain configurations.

The general scheme of things is that the sun shines through the glazing, where it heats up the thermal mass of the wall behind. The wall, being of a construction that has a high thermal mass (for example masonry or concrete) transfers some of the heat energy to the air in the gap between the glazing and the wall as it heats up.

A convection current is set up. If you are familiar with heat and the way it affects air, you

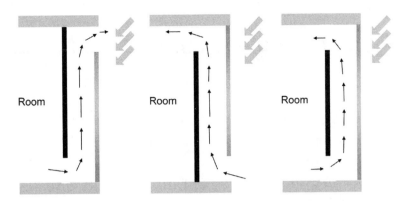

Figure 5-2 *Trombe wall modes of operation.*

Passive evaporative techniques

When water evaporates, it takes with it energy. We can exploit this phenomenon to cool buildings. These techniques all require water, which might not be possible in some hotter countries where water availability is limited. Also, it must not be forgotten that there is a requirement for energy to pump the water to the top of the building. This energy must be provided in a sustainable manner.

Roof sprays

Spraying the roof with a fine mist of water is one way to keep the roof wet and permit evaporative cooling. The roof must be suitably coated to prevent water ingress, which could damage the fabric of the building.

Roof ponds

A roof pond is one way of providing a large body of water which can be evaporated, taking heat with it as it leaves the roof.

Active solar cooling

Active solar cooling is a little bit more involved than passive solar cooling. In active solar cooling, we use a thermally driven process of some sort in order to cool our buildings, rather than air conditioning which is thirsty for electricity. Of course, as we have seen, we can easily generate heat using solar methods.

To understand how solar cooling differs from conventional refrigeration methods, let's compare the two and look at similarities and differences.

In a conventional refrigeration setup, a refrigerant—a substance that readily evaporates at a low temperature—is compressed, which causes it to become liquid. This compression is usually driven by an electric motor—using valuable watts in the process. The refrigerant is then allowed to expand—to do this it requires heat, which it gains from the material under refrigeration. As the heat transfers from the material to be refrigerated to the refrigerant, the refrigerant expands. It must then be compressed and forced around the loop again! This cycle continues indefinitely—no refrigerant should escape from the system.

Our system works in a slightly different way. The refrigerant is kept "locked up" in a material which soaks up refrigerant like a sponge soaks up water. As we heat this material, the refrigerant is liberated from it, turning into a liquid as it condenses. This liquid will readily evaporate again—it is encouraged to do this by the absorbent material which tries to "suck" the refrigerant back once it has cooled. As the refrigerant shuttles back to the absorbent material it takes heat with it. This shuttling back and forth continues—so the process is a bit more like a train going back and forth in a straight line, than a train continually circling in a loop.

Project 7: Solar-Powered Ice-Maker

I am grateful to Jaroslav Vanek, Mark "Moth" Green and Steven Vanek for the information on how to make a solar-powered ice-maker. This design was originally published in *Home Power* magazine, Issue no. 53.

The original article can be downloaded from the *Home Power* website at:

homepower.com/files/solarice.pdf

You will need

- Four sheets galvanized metal, 26 ga.
- 3 in. black iron pipe, 21 ft length
- 120 sq ft mirror plastic
- 2 ¼ in. stainless steel valves
- Evaporator/tank (4 in. pipe)
- Freezer box (free if scavenged)
- 4 ft × 8 ft sheet ¾ in. plywood
- six 2 × 4 timbers, 10 ft long
- Miscellaneous ¼ in. plumbing
- Two 3 in. caps
- 1 ¼ in. black iron pipe, 21 ft length
- Four 78 in. long 1 ½ in. angle iron supports
- 15 lb ammonia
- 10 lb calcium chloride

This design is for an ice-maker which will produce about 10 lb of ice in a single cycle. It uses the evaporation and condensation of ammonia as a refrigerant. If you remember in the explanation above, I mentioned that we needed a refrigerant and an absorber for this type of cooler to work. Well, the ammonia is our refrigerant, and we use a salt—calcium chloride—as the absorber. You might have seen small gas fridges often used in caravans and RVs which can be powered by propane—these also generally use ammonia as a refrigerant—however, they tend to use water as the absorbing medium.

Construction and assembly

The first item to be assembled is the solar collector pipe. This is made from the length of black iron pipe. First of all you should cut a foot off the end, as we will need this for the ammonia storage tank. The pipe should be capped at the ends with the 3 in. black caps, but before you do this, you need to drill one of the caps to accept a ¼ in. nipple and a coupling for the rest of the plumbing. The collector can now be filled with the calcium chloride salt which will act as the absorber. The caps can now be secured firmly in place. Whatever method you choose, you should ensure that the joint is capable of withstanding pressure—as when the ammonia is produced it will be hot and anywhere near two hundred pounds per square inch pressure.

You next need to form a condenser coil and tank. The tank is easy—take a standard 55 gallon drum, and slice it in half. This will give you a nice container to pump full of water.

Now a word about gravity—there are no pumps in this system to make the working fluid go up and down, so you need to think of other ways of doing this. Mount the condenser coil high up, above the level of the solar collector. Now have the pipe from the collector running to the top of the coil in the tank. The pipe from the bottom of the coil to the storage tank should be as straight a run downhill as possible, try to eliminate any bends or kinks if at all possible (Figures 5-3 and 5-4).

In the collector made by Jaroslav Vanek, Mark "Moth" Green and Steven Vanek, the steel pipe of the collector was supported from the ground by two sturdy uprights. The solar collector was then

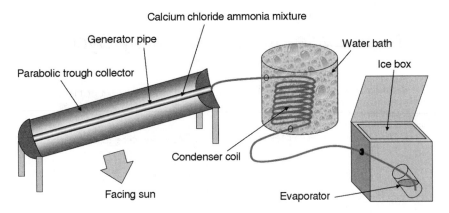

Figure 5-3   *Solar cooler layout.*

suspended from this using U bolts. This allowed the collector to be moved to accommodate seasonal variations.

How does the ice-maker work?

The ice-maker works on a cycle—during the daytime ammonia is evaporated from the pipe at the focal point of the parabolic mirrors. This is because the sun shines on the collector which is painted black to absorb the solar energy—this collector heats up, driving the ammonia from the salt inside.

At night, the salt cools and absorbs the ammonia, as it does this, it sucks it back through the collector. As it evaporates from the storage vessel, it takes heat with it.

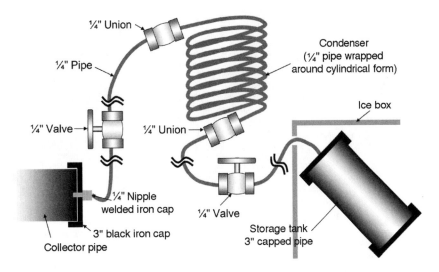

Figure 5-4   *Solar cooler plumbing details.*

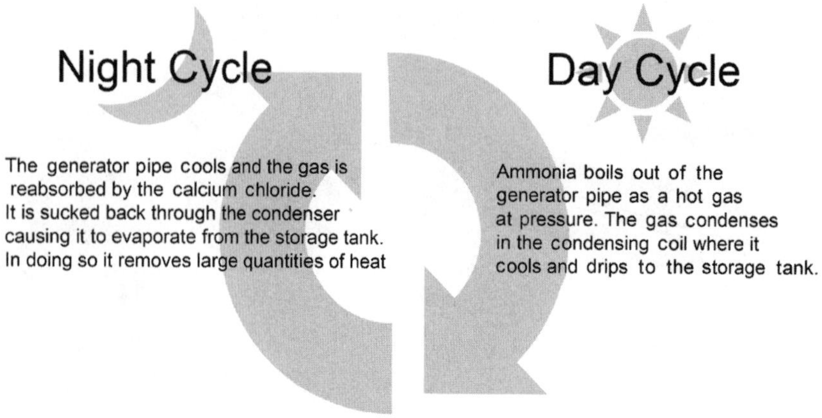

Night Cycle

The generator pipe cools and the gas is
reabsorbed by the calcium chloride.
It is sucked back through the condenser
causing it to evaporate from the storage tank.
In doing so it removes large quantities of heat

Day Cycle

Ammonia boils out of the
generator pipe as a hot gas
at pressure. The gas condenses
in the condensing coil where it
cools and drips to the storage tank.

Figure 5-5 *The solar cooler cycle.*

Figure 5-5 illustrates this cycle.

Note

This design has a number of strengths which
make it robust and reliable in operation. One of
those strengths is that the design has an absolute
minimum of moving parts. The only things that
actively move are the valves—and even these are
operated infrequently. The elimination of moving
parts makes this design very efficient.

Useful addresses

The following addresses may be useful if you wish to
make further enquiries about this design of ice-maker:

S.T.E.V.E.N. Foundation,
414 Triphammer Rd.
Ithaca,
NY 14850
U.S.A.

SIFAT,
Route 1, Box D-14
Lineville,
AL 36266
U.S.A.

Chapter 6

Solar Cooking

Why cook using the sun?

Solar cooking is a great alternative to conventional cooking—rather than burning fuel and producing carbon dioxide emissions, or using precious electricity, solar cooking harnesses the natural energy available from the sun! It is a great social activity on a sunny day—barbeques are just sooooo yesterday—solar cooking is where it is at! No fumbling with matches and firelighters, no choking on smoke, no burnt sausages! Just hope that the clouds don't come!

Although you won't see any T.V. chefs preparing meals on a solar cooker, it doesn't mean that they aren't any good—it just means that T.V. chefs lack technological imagination.

There are *lots* of different designs of solar cooker all suited to different applications—all rely on similar principles—concentrating the sun's energy into a small area and then trying to retain the heat.

Solar cooking solutions are elegant in their simplicity and as such are suited to developing world applications (Figure 6-1). Many countries do not have the developed infrastructure that we have in the West for distributing energy. As a result, a hot, cooked meal is hard to come by—as fuel may be scarce.

Think of it like this—the developed world is already using a massive amount of energy to cook food—with large nations like India and China growing and developing, our energy will run out sooner rather than later if everybody wants to live a western lifestyle—but why even should we in the West want to live a western lifestyle when things like solar cooking can be so much fun—and achieve the same ends that conventional cooking does, heating a food product.

All of the projects in this chapter can be built *very* cheaply and are ideal for a fun summer's day!

At the end of this chapter, I have put together a collection of links to various different types of solar cooker plans that are out there on the web—all have different strengths and weaknesses and are suited to different applications, from designs that will just about cook a frankfurter, to large cookers that can be used for community catering!

Figure 6-1 *A solar cooker being used in the developing world.* Image courtesy Tom Sponheim.

Project 8: Build a Solar Hot Dog Cooker

You will need

- Small photovoltaic cell
- Solar motor
- Plywood
- Framing
- Screws
- Flexible acrylic mirror
- Elastic band and pulley wheel

    or

- Small plastic worm gear and large plastic gear to match

Tools

- Bandsaw
- Drill
- Router

This list of components is for the deluxe version—the automatic hot dog turner is a cool novelty, but not essential as it is just as easy to turn by hand! If you want to make a simpler cheaper version, substitute the plywood for card, the plastic mirror for tin foil, and rather than have a motor turn your hot dog on a skewer, just provide support for the skewer and do it by hand!

First of all we need to construct the parabolic mirror. The parabolic mirror collects all of the solar energy and focuses it onto our hot dog. You can read more about this effect in the chapter on Solar Collectors (Chapter 8).

Have you ever tried bending a glass mirror? It doesn't work, the least you will end up with is seven years bad luck, and at worse you could end up with bloody hands.

Here we have a couple of options, the "quality-built-to-last" option, and the "cheapskate" option. The quality-built-to-last option involves buying some flexible acrylic mirror from the internet. This is often seen on auction sites in large panels as people use it for interior design, and can sometimes be found in garden supply stores as people put it at strategic locations in their garden in order for it to appear bigger.

The acrylic mirror can be bent gently without fear of breaking, and also has the advantage that it can be drilled relatively easily without fear of splintering.

If you are opting for the cheap option, you can get away with using some corrugated cardboard, covered in glue, followed by tin foil. Note that your collector efficiency will not be as good, as the surface is not as reflective. However, for a demonstration it works to a degree.

Next, you need to construct a support for the mirror—if you are building to last then use plywood and framing. A router might be useful to machine a groove which will support the mirror, if you are making the cheaper cardboard version, you just need to cut some flaps to support the "mirror."

Now we come to the drive mechanism. We have a couple of options here (Figures 6-2 and 6-3). You can just provide a simple support for a skewer and turn your hot dog by hand.

Alternatively, if you are feeling really adventurous, you can construct a solar motor, which will turn the skewer of your hot dog automatically!

You will need

- 1381 IC
- 2N3904 transistor
- 2N3906 transistor
- 3300 μF capacitor
- 2.2 k resistor

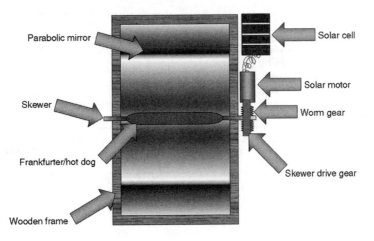

Figure 6-2 *Solar hot dog cooker.*

- Solar cell
- High efficiency motor

The circuit for the solar motor is shown in Figure 6-4, it is a simple circuit, and can easily be assembled on stripboard. Once you have constructed the solar motor driver circuit, you are going to need to mechanically couple the motor to the skewer. You may find if your motor is powerful enough, that you can directly drive the skewer—it would be worth investigating bearings to ensure that your skewer turns as freely as possible with a minimum of resistance. If you find your motor struggles to turn the skewer, then use a "worm drive" to reduce the speed of the motor—while increasing torque.

Online resources

Solar hot dog cookers on the web:

www.motherearthnews.com/Green_Home_
Building/1978_March_April/Mother_s_Solar_
Powered_Hot_Dog_Cooker

www.pitsco.com/the_cause/cause3inv.htm

www.energyquest.ca.gov/projects/solardogs.html

sci-toys.com/scitoys/scitoys/light/solar_
hotdog_cooker.html

www.reachoutmichigan.org/funexperiments/
agesubject/lessons/energy/solardogs.html

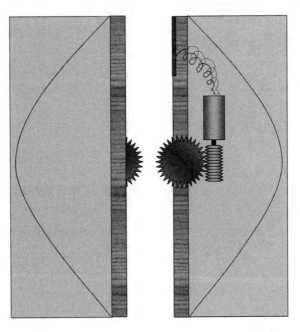

Figure 6-3 *Drive mechanism.*

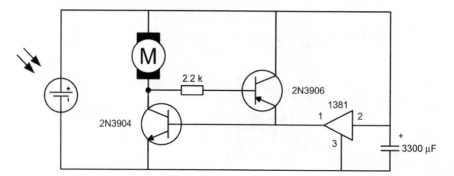

Figure 6-4 *Solar motor circuit.*

Project 9: Build a Solar Marshmallow Melter

You will need

- Marshmallows
- Large Fresnel lens
- Tin foil
- Skewers or a toasting fork

In this project, we are going to collect the sun's energy from a large area, and focus it to a point in order to create localized heating.

One way of collecting the sun's energy from a large area is to use mirrors. We have already explored this in the "solar hot dog cooker." You will read more about concentrating solar energy in the chapter on Solar Collectors (Chapter 8).

In order to perform this experiment, we are going to need a Fresnel lens—again see the chapter on Solar Collectors for an explanation on how these work.

Put the marshmallow on a skewer, and rest it on the sheet of tin foil. We are going to use the Fresnel lens, to focus the sun's rays onto the marshmallow. When you look out of your window, there is no magnification or reduction of the image—the glass does not act like a lens; however, you will notice when looking through the Fresnel lens (not at the sun!) the image appears much bigger and magnified. Why is this? If you look closely, you will see a series of concentric circles in the Fresnel lens. Now think of a magnifying glass—it is round and circular, and "bulges" in the middle. If we look at the glass from side-on, we can see that both sides of the lens are curved—but there is also a lot of glass in the middle! A Fresnel lens "removes" some of the glass from the middle, and flattens the lens onto a sheet. Each little concentric ring that you see on the flat Fresnel sheet, is a section of lens curve.

Look at where the sun is in the sky, and hold the Fresnel lens perpendicular to an imaginary line between the sun and your marshmallow. Move the lens to and fro along this line, and observe how the focused beam of solar energy changes on your marshmallow. After a little bit of time, focusing the sun onto the marshmallow, you should see the candy begin to toast! No fire required—just the power of the sun!

Online resources

Marshmallow melting web pages!

worldwatts.com/marshmallows/solar_roaster.html

www.altenergyhobbystore.com/marshmallow%20roaster.htm

bellnetweb.brc.tamus.edu/res_grid/cuecee05.htm

You will need

- Eggs!
- Drop of oil
- Hot sunny day

Tools

- Black tarmac driveway
- Frying pan

Sometimes, on a hot sunny day, the black tarmac can almost seem painful to walk on barefoot as it is so hot. If you keep moving, your feet feel fine; however, if you stand in the same place for the same time, your feet feel very uncomfortable. This is because the tarmac road surface has the ability to act as a thermal mass and store heat. If you were to stand on say a flimsy piece of black card that had been left in the sun, it would feel warm to the touch; however, you would find that as soon as you stood on it, the heat would be quickly dissipated—the card doesn't have the ability to store the heat. So, if we want to cook an egg on a sunny day . . .

Take a peek at Figure 6-5 for the ridiculously simple method.

Take a frying pan, put it on a black tarmac surface on a hot sunny day, put a drop of oil in the pan and cover the frying pan for a while with a sheet of glass. The pan is black, the tarmac is black and so will have absorbed the sun's energy. All of this heat via one process or another will transfer to the oil, and pretty soon you should have hot oil. Now crack an egg, and you will find that it cooks—once again cover the pan with a sheet of glass. Of course, this trick requires the right sort of day—don't expect fried egg on a cloudy day in

Alaska! But if your climate permits, this is a nice trick! If there is not as much sun as you would like, try using reflectors to aim more solar energy onto your pan!

In fact, with simple solar cooking, I have even heard of people baking cookies in their car by simply putting a black baking tray with cookie dough on their dashboard, and parking the car in a sunny setting with the windows up. They then return to the car at lunch to find a tray of cookies and a "bakery fresh" smelling car. It sure beats a Magic Tree for in-car air freshening!

Online resources

The following link is a great solar cooker site written specifically for younger kids.
pbskids.org/zoom/activities/sci/solarcookers.html

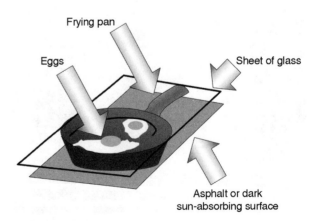

Figure 6-5 *Solar egg frying.*

You will need

- Sheet of thin MDF
- Sheet of flexible mirror plastic
- Sheet of thin polystyrene
- Veneer panel pins

Tools

- Bandsaw
- Pin hammer
- Sharp knife/scalpel
- Angle marking gauge

This solar cooker is a very simple project to construct—we will be harnessing the sun's energy from a relatively wide area and concentrating it to a smaller area using mirrors (read more in Chapter 8 about this). The area which we will concentrate it into will be lined with polystyrene to keep in the heat.

Construct a box for your cooker out of MDF. I find small veneer pins to be very useful as they can be hammered neatly into the end grain of thin MDF without splitting the wood. For this application they are perfectly strong enough. When you have finished the box it should look something like Figure 6-6.

Now you need to line the box with polystyrene, this will prevent the heat from escaping. The lined box will look like Figure 6-7.

Now measure the size of the cube inside the lined polystyrene box. You should cut the mirror plastic to this size, and further line the box with it. Duck Tape is more than ideal for making good all of the joints and securing things into place.

We now need to cut the mirrored reflectors. Cut a strip of mirror plastic about two feet wide on the bandsaw. Now, using an angle marking gauge, mark from the long side of the mirror to the very corner of the mirror, a line which makes an angle of 67°, forming a right-angled triangle in the scrap piece of plastic. You now need to mark out a series of trapeziums along this length of mirror, where

Figure 6-6 *The box constructed from MDF.*

Figure 6-7 *The box lined with polystyrene.*

Project 11: Build a Solar Cooker

Figure 6-8 *The mirrored reflectors cut ready.*

Figure 6-9 *The solar cooker ready and complete.*

the shortest side is equal to the length of the inside of the box cooker (Figure 6-8).

Now take the mirrored reflectors, and on the nonreflective side, use Duck Tape to join them together to form the reflector which will sit on the top. Using Duck Tape allows you to make flexible hinges, which allow the reflector to be folded and stored out of the way.

When the cooker is finished it will look like Figure 6-9. It is now ready for cooking!

Project 12: Build a Solar Camping Stove

You will need

- Five sheets of A4 or U.S. letter size cardboard
- Tin foil
- Glue
- Adhesive tape

Tools

- Scissors

This is an incredibly simple construction for a solar camping stove.

Simply, take five sheets of cardboard—three of them should be joined together by their long edges, the other two should be joined up by their short edges. Make the joint using adhesive tape so that it is flexible.

Now cover the two pieces you are left with in tin foil. Use glue to secure the foil.

And that is it! Now all it comes to is setting up your stove.

Determine which way the sun is facing, and orient the three panels so that they all face the sun, with the outer two tilted slightly inwards. Now take the two sheets, one will sit on the ground—the food stands on top of this. The second sheet should be slightly tilted up toward the can so that any overspill light which misses the food is reflected back onto it.

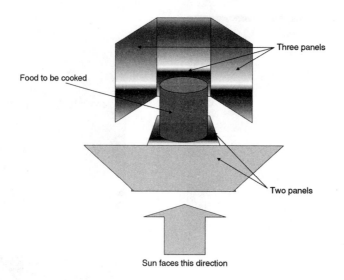

Three panels

Food to be cooked

Two panels

Sun faces this direction

Figure 6-10   *The set-up solar stove.*

The beauty of this design is that it is very simple, can be assembled quickly, and fits into the space of a few sheets of cardboard in your backpack.

The set-up cooker is illustrated in Figure 6-10.

Online resources

There are a lot of folk out there who swear by solar cooking. All sorts of people have formulated different designs of solar cooker. While I have presented a few designs that have worked for me, there are *many, many* other different types of design for different applications. I highly recommend that you browse some of the following links in order to find the solar cooker that is right for you.

The Tracking Solar Box Cooker
solarcooking.org/Cookerbo.pdf
Fresnel Reflector Cooker
www.sunspot.org.uk/ed/
The Reflective Solar Box Cooker
solarcooking.org/newpanel.htm
Collapsible Solar Box Cooker
solarcooking.org/collapsible-box.htm
The Bernard Solar Panel Cooker
solarcooking.org/spc.htm

The Easy Lid Solar Cooker
solarcooking.org/easylid.htm
The Minimum Solar Box Cooker
solarcooking.org/minimum.htm
Heaven's Flame Solar Cooker
www.backwoodshome.com/articles/
radabaugh30.html
The Cooking Family Solar Panel
solarcooking.org/cookit.htm
Inclined Box Type Solar Cooker
solarcooking.org/inclined-box-cooker.htm
Sun Pan Solar Cooker
www.sungravity.com/sunpan_overview.html
Nelpa Solar Cooker
solarcooking.org/nelpa.htm
Pentagon Star Coooker
solarcooking.org/PentagonStar.htm
Dual Setting Panel Cooker
solarcooking.org/DSPC-Cooker.htm
A cardboard and tinfoil cooker with two heat settings.
Solar Funnel Cooking
solarcooking.org/funnel.htm
The Tire Cooker
solarcooking.org/tire_eng.htm
A solar cooker made from a recycled tire!

Continued

Online resources—cont'd

Parvarti Solar Cooker

www.angelfire.com/80s/shobhapardeshi/twelvesided.html

Windshield Shade Solar Cooker

solarcooking.org/windshield-cooker.htm

A solar cooker made from an old car windscreen reflector shade

Double Angle Twelve Sided Cooker

solarcooking.org/DATS.htm

A simple cooker design from cardboard and tin foil

Parabolic Solar Cooker

www.sunspot.org.uk/Prototypes.htm

A solar cooker with an aluminum reflector and card base

Solar Bottle Pasteurizer

solarcooking.org/soda-bottle-pasteurizer.htm

A pasteurizing device powered by the sun, and made from recycled materials

Solar Water Pasteurizer

solarcooking.org/spasteur.htm

Solar Chimney Dehydrator

www.littlecolorado.org/solar.htm

Simple plans to build a food drying device powered by the sun

Solar Cooking in the Peruvian Andes

www.sunspot.org.uk/Solar.htm

Solar cooking in the developing world

Solar cooking recipes

Potatoes

For a start, cooking potatoes with a solar cooker differs a bit from cooking them in a campfire, which you are probably used to, because if you wrap them in shiny reflective tin foil, the solar energy which you have gone to painstaking ends to concentrate onto the potato will simply be reflected!

Baked potatoes

This is a really nice cartoon about cooking potatoes in the sun.

www.hunkinsexperiments.com/pages/potatoes.htm

Brewing tea

If you want to brew tea in a solar cooker, you can't expect to get boiling water and then make your tea conventionally—instead take a jar and a couple of tea bags, put the tea bags in the jar along with some clean water (which you might have even got from your solar distilling apparatus!).

Soups

Soups are really easy to cook in a solar cooker. Furthermore, they are particularly forgiving if the amount of sunlight is suboptimal, as warmish vegetable soup is quite acceptable whereas rawish not fully cooked chicken is totally unacceptable!

Nachos

Everyone loves Nachos! So why not take a bag, spread them in a bowl and cover with grated cheese. Then place the bowl in your solar cooker to melt the cheese and give you toasty hot nachos!

Bread

Take some old baked bean tins and paint them black—you now have the perfect can for cooking bread!

To cook some simple French bread you will need a packet of baker's yeast, a tablespoon of sugar and a tablespoon of salt, five cups of white flour and a couple of cups of water.

Dissolve the yeast in one cup of slightly warm water. Sift all of the dry matter into a clean bowl, stir in the yeast—water mix, add the water from the second cup in small amounts until the dough is sticky. Grease a baked bean can which has been painted black, being careful of any sharp edges, add the bread mixture and leave it in your solar oven.

Solar cooking tips

In many campsites and caravan parks, open fires are banned because of the mess they produce and the smoke which can be unpleasant for other visitors—so while everyone else has run out of gas in their cylinder, or is eating cold raw food, now would be a great time to crack open the solar cooker and make the rest of the campsite jealous!

You really want to cook on days when the sky is clear and the sun can easily be seen—on a cloudy day, cooking will be painfully slow.

One of the great things about solar cooking is that you can prepare everything in advance, leave it in your solar cooker, and when you return everything is cooked ready to eat—whereas your accomplices cooking with traditional methods still have to muck about and cook their food!

Also—think like this—if you are cooking using conventional energy inside a home that is air conditioned, for every kWh of energy you input to your cooker, your air conditioning will use about another three trying to remove that heat from your home!

Chapter 7

Solar Stills

Water—a precious resource

The former World Bank Vice President Ismail Serageldin, said that "the next world war will be over water."

At first look, this statement seems almost non-sensical, we are surrounded by water, it falls from the skies and runs through our streams and rivers; however, not all of the world enjoys such plentiful access to water as we do in the developed world.

In much of the developing world the land is arid, and clean drinking water can often mean a walk for tens of miles. This problem is exacerbated by heavy industry building factories which extract what little water there is.

Our water is constantly recycled by the natural environment, it follows a pattern called the hydrological cycle, which can be very simply represented by Figure 7-1.

Water evaporates from the earth, plants, animals, and people, is carried far up into the sky where it condenses to form clouds—then it precipitates back to earth in the form of rain.

This has a purifying effect on the rainwater, as when the water evaporates, contaminants are left behind—or at least this used to be the case—sulfur dioxide and other nastiness in the air from human activity can be collected by the rain as it precipitates, with the effect that when it lands on the earth, it is acidic. This can cause problems for plants and alkaline rocks, which are damaged by the acid content of the rain.

A solar still effectively creates the hydrological cycle in miniature in an enclosed volume. The idea is that by evaporating water, all of the bacteria, salts and other contaminants are left behind, with the precipitate being pure, drinkable water.

Even seawater can be desalinated using this process.

There are a number of advantages to solar distillation:

- Free energy
- No prime movers required

History of the solar still

Solar stills are an old, tried and tested technology—the earliest record of a solar still being used is in 1551, when Arab alchemists used one to purify water.

Mouchot, whose name also springs up a couple of other times in this book also worked with solar distillation around 1869.

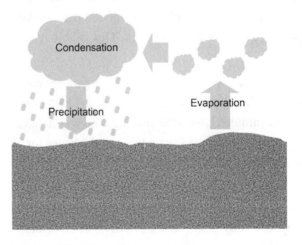

Figure 7-1 *The hydrological cycle.*

The first solar still, in the sense that we would recognize it now, was built in the mining community of Las Salinas in 1872, in the area which is now northern Chile. It was created by a Swedish technologist by the name of Charles Wilson. The plant was massive, about 4,700 square meters—quite an engineering feat for the time.

The plant produced in excess of 6,000 gallons of water a day.

The plant was effective and produced water well into the 20th century until it was finally closed in 1912. All that remains now are shards of glass and salt deposits in the area where the stills were originally constructed.

Project 13: Build a Window-Sill Demonstration Solar Still

You will need:

- Pint glass
- Egg cup
- Cling film/Saran wrap
- Sellotape
- Penny
- Tea bag

Tools

- Scissors
- Kettle

This is a demonstration of how solar still technology works. It works great as a science fair demonstration piece, and is of a size that you can quickly put it together in a few minutes. This is how it works.

First of all, we are going to make our water brackish. The best way to do this is to put the kettle on! A few minutes later, after making a brew you have black tea. Allow the teabag to sit in the water for some time until the water is quite "muddy."

Now put your egg cup in the bottom of the pint glass, and while holding the egg cup out of the way, carefully pour the tea into the bottom of the "still" making sure not to get any in the egg cup.

Now take some clear plastic such as cling film/saran wrap and stretch it over the top of the pint glass. You might want to anchor it around the perimeter of the glass using a little bit of Sellotape just to make sure.

You will want to stick your finger into the plastic in order to stretch it a little bit and create a dip above the egg cup. Be careful not to stick your finger through!

You might want to put a little weight, such as a small coin, here in order to preserve the dip. Your whole assembly should now look something like Figure 7-2.

Figure 7-2 *Demonstration solar still.*

Put the glass on a south-facing window sill and leave it for a couple of days. After some time, the plastic on top of the still will look something like Figure 7-3.

This water should taste clear and pure, not "brackish" (i.e. strong tea!).

Now you have proved the operation of the solar still!

Figure 7-3 *Precipitated water in the demonstration still.*

Project 14: Build a Pit-Type Solar Still

You will need:

- Polythene sheet
- Cup
- Tube
- Rocks

Tools

- Spade

This type of solar still is ideal if you are camping in a hot climate or stuck in the desert and you need to extract some clean drinking water.

Warning

On any camping expedition, remember to take sufficient water with you for the amount of people and time you will be away. This type of still should only be used as a demonstration or in emergencies, and does not provide a consistent reliable method for providing water for your travels, beyond basic, emergency needs.

First of all, you will need to dig a hole with a spade. In this hole, you can place green plants, cacti, pots of brackish water or anything else you can gather that could potentially be water bearing.

In the middle of this hole, you need to put a small cup, bowl or receptacle for water. A tube runs from this receptacle to outside the hole. Water can be extracted using this tube without having to upset the solar still or dismantle it.

Over the top of the still you need to put a clear polythene sheet. This should be weighted down around the edges using rocks and stones. A small, light weight should be put in the center dip in order to let the water settle to a point for collection. This is shown in Figure 7-4.

Water will precipitate and collect in the receiving vessel over time. In order to collect the water, just give a little suck on the pipe as shown in Figure 7-5 and pure water will come from the still.

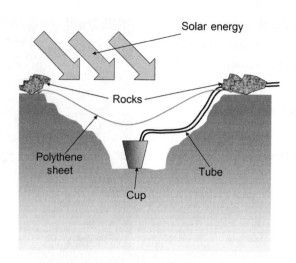

Figure 7-4 *Diagram of a pit solar still.*

Figure 7-5 *A solar still in operation.* Image courtesy © U.S. Department of Agriculture—Agricultural Research Service.

Project 15: Build a Solar Basin Still

You will need:

- Plywood/oriented strand board
- Framing
- Screws
- Glazing (glass/polycarbonate)
- Metal U-strip
- Black silicone
- Low-profile guttering
- Low-profile guttering end pieces
- Tube
- Two stop cock valves

Tools

- Jigsaw
- Screwdriver
- Squeegee

This project is scaleable depending on your requirements for water, which is why no specific measurements are presented.

First of all, you will need to calculate your water needs. Solar stills can generally produce around a gallon of water per 8 square feet, this is around four liters per square meter. This assumes that your collector receives 5 hours of good sunlight per day. Obviously the performance of your still will be highly variable, depending on the amount of sun your collector receives.

You need to construct a wooden box from plywood or oriented strand board, with gently sloping sides. This is well within the capability of someone with even modest carpentry skills.

At a position near the tallest side of the box you will need to drill a hole and insert a pipe with a valve that can be opened and shut, to allow you to introduce brackish water to be purified.

Then, take a squeegee and some black silicone. You need to spread this mixture on the bottom face of the wooden box so that it gets a thin uniform coat. Less important are the sides, but you should ensure that by the time you are finished, the inside of the box is fully lined with silicone.

At the front of the still, that is to say on the shortest side of the box, you need to make a small gutter. This gutter will serve to collect your purified water which will run down by the force of gravity from your glazing. You need to make this gutter out of a waterproof material. The low-profile guttering sold for sheds and outbuildings is ideal. A hole needs to be drilled in the side of the frame of your still, and a pipe introduced to allow you to siphon off the clean water.

The silicone has two functions. First of all, it acts as a black collector surface, absorbing radiation and creating heat. But secondly, it protects your wood by making the enclosure waterproof.

On top of this sealed box you need to put a sheet of glazing. This needs to be sealed around the edges with frame sealant to ensure a good watertight fit.

The brackish water should never be allowed to rise above the level of the guttering, as it would contaminate the clean water. The whole solar still is illustrated diagrammatically in Figure 7-6.

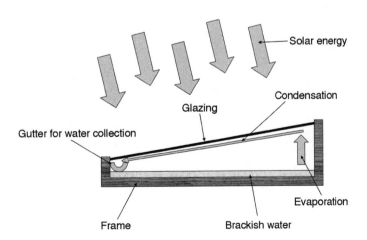

Figure 7-6 *Diagram of the basin type still.*

Chapter 8

Solar Collectors

The sun provides an abundance of energy over a wide area; however, often our solar devices are fairly small, and so receive little solar energy. So—what if we could take the solar energy from over a wide area, and concentrate it into a smaller area? This makes a lot of sense, because it means that the small area receives a much higher amount of solar radiation.

So what can solar collectors actually do?

Actually, the sun has quite a phenomenal power—when concentrated into a small area, its power is truly extraordinary. If you were a ghastly child you might have burnt ants using a magnifying glass—well what goes around comes around: remember that when a large ethereal figure holds a magnifying glass over you. One of my memories of junior school was gathering in a corner of the playground where a group of children were concentrating the sun onto some logs covered in tar and making smoke. Although we did not know it then, we had made a solar collector.

The chances are you're getting tired of reading this, but "this is not a new concept," in fact, the Greek's purportedly had a "weapon of mass destruction," that harnessed the power of the sun to set fire to enemy boats.

Archimedes—you may have heard of him—he found a few things out, like the concept of the Archimedes screw and the theory of displacement. Anyway, it is fabled that he had a weapon that was created out of mirror-like bronze that he could use as a death ray—this ray essentially reflected concentrated sunlight!

In the book *Epitome ton Istorion*, John Zonaras wrote: "At last in an incredible manner he burned up the whole Roman fleet. For by tilting a kind of mirror toward the sun he concentrated the sun's beam upon it; and owing to the thickness and smoothness of the mirror he ignited the air from this beam and kindled a great flame, the whole of which he directed upon the ships that lay at anchor in the path of the fire, until he consumed them all."

This deadly weapon was allegedly used in the siege of Syracuse in 212 BC—like I said, the idea is *old*!

So this is what MIT did . . .

First they got *loads* of students on the 2.009 course, *loads* of chairs to act as stands and *loads* of mirrors (Figure 8-1). Being MIT, they got the cash for this kinda stuff!

Next they lined all the mirrors up so that the sun's energy was concentrated onto the model of the hull of a boat—voila!—or should that be Eureka? Flames! (Figure 8-2)

Here we can see the *serious* damage done by the flames to the wood (Figure 8-3)! With a larger mirror area this could have been a formidable weapon!

In Figure 8-4 we see how MIT used a similar technique to the one you will use in the next project—cover each mirror up with paper, line each one up individually by removing the piece of paper and adjusting the mirror. And then, when they are all lined up, remove all the bits of paper as fast as you can without disturbing the mirrors!

And as ever, with every serious piece of technological investigation, there is the back of the paper bag calculation (Figure 8-5).

Now it's your go!

Figure 8-1   *Students, chairs and mirrors!* Image courtesy Massachusetts Institute of Technology.

Figure 8-2   *The boat catches fire—Archimedes was right.* Image courtesy Massachusetts Institute of Technology.

Figure 8-3 *The burnt hull.* Image courtesy Massachusetts Institute of Technology.

Figure 8-4 *Lining up the mirrors.* Image courtesy Massachusetts Institute of Technology.

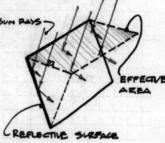

WOOD PLACED ON AN ELECTRIC RANGE ELEMENT BURNS
A LARGE RANGE ELEMENT USES $\simeq$ 1500 W

$\therefore \simeq$ 1500 W/ft² SHOULD BURN WOOD

SOLAR INSOLATION IS $\simeq$ 1000 W/m²

$\therefore \simeq$ 1.5 m² CONCENTRATED ON 1 FT² IDEALLY WOULD BURN WOOD

GUESS THAT GEOMETRY IS SUCH THAT EFFECTIVE COLLECTING AREA IS REDUCED BY ½.

▶ ESTIMATE: 3m² COLLECTOR FOCUSED ON 1 SQUARE FOOT SHOULD IGNITE WOOD

SUN RAYS
EFFECTIVE AREA
REFLECTIVE SURFACE

Figure 8-5 *Working it all out!* Image courtesy Massachusetts Institute of Technology.

Project 16: Build Your Own "Solar Death Ray"

You will need:

- Sheet of MDF
- Sheet of flexible mirror acrylic
- 72 long self-tapping screws

Optional

- Silicone sealant

Tools

- Drill bit
- Hand/cordless drill
- Glue gun and sticks
- Ruler
- Set square
- Bandsaw

Optional

- Mastic gun

Warning

I have used acrylic mirror in this project because it is very easy to work with, and can be cut easily using a band saw; however, there is nothing to stop you using a glass mirror if it is available and you have the correct tools to cut it and work with it—my only advice is it will be harder to work with and much more fragile.

OK, so you have finally decided—the time is nigh to melt your little brother. While he might be hard to melt, you can certainly singe him with this modular solar death ray!

Don't worry—you won't need lots of chairs and big A4 mirrors like the guys at MIT! Instead, this modular death ray relies on little tiles which are cut from plastic mirror.

The plan is really simple—you build the death ray a tile at a time. One tile is good to experiment with, but once you become more confident and want to expand, you can simply add more tiles!

To begin with, I recommend that you cut yourself a piece of MDF that is 36 cm square, although please bear in mind that this measurement is wholly arbitrary.

Now using a ruler and set square, divide the sheet into a matrix of six squares by six squares. This will give you thirty-six equal squares 6 cm square. Now, using the ruler and set square, draw a line 1 cm either side of each line making up the squares. This will leave you with a sheet that does not look dissimilar to Figure 8-6.

You are now going to drill holes for the screws that will support the mirrors. You will need to select a drill that is slightly smaller than the screw that you are going to drill the hole for. However, please note that the screw does not need to be a tight fit in the hole, as it would be if you were joining two pieces of wood. Instead, the screw is only going to be used for light adjustment, so the screw can be a relatively slack fit in the hole.

Looking at your board of squares, you are going to be drilling two holes in each 6 cm square. The holes will be at the top left and bottom right, where the lines cross to form the smaller square inside each square. Sounds confusing, well, take a look at the furnace drilling diagram (Figure 8-7), which shows where to drill in each square.

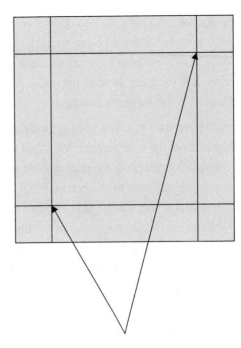

Figure 8-7 *Furnace drilling diagram.*

Once you have drilled all 72 holes, you are going to need to think about getting those screws in place. This is a *really* tedious job, so either ask a younger sibling, or failing that, if you are an only child you might like to consider investing in an electric screwdriver—the lazy man's way out.

You want to put the screws in so that they just protrude from the other side a little way (Figure 8-8).

Now is a good time to take your acrylic sheet of mirror, and, on a bandsaw, cut 36 identical

Figure 8-6 *Sheet of MDF marked out.*

Figure 8-8 *Solar furnace with the screws in place.*

6 × 6 cm squares. An easy way to do this is to set the gate on your band saw to 6 cm from the blade. Take a couple of cuts from your mirror, to give you 6 cm strips, and then cut these strips into squares using the gate at the same measurement.

Once you have done this, you need to fix the mirrors to the base plate of the solar furnace. You need to pick a corner, which is different to the ones that the screws are positioned in, and stick to this corner. What you will be doing is applying a large glob of either glue or silicone sealant, into which the corner of the mirror tile is immersed. The other two corners are supported on screws, which permit adjustment of the tile's angle relative to the base board (Figure 8-9).

When all the mirrors are in place, stick a small removable piece of paper, for example a Post-It note to each of the mirrors.

Set your collector up so that it faces the sun. Remove one of the pieces of paper in one of the corner mirrors—notice where the light forms a bright patch and set up an object to be heated or piece of wood there. Draw an X where there is a bright patch.

Now, one by one, using the screws for adjustment, you can change the angle of each individual mirror. Cover and uncover mirrors one by one using the Post-It notes—you will need to work quickly as you will find that the sun is constantly changing position.

Figure 8-9 *Board with mirrors stuck on starting to take shape!*

Eventually you will find that you can focus all of the mirrors onto a single point—this concentrated energy can be used for cooking, heating, or experiments (burning things!).

Parabolic dish concentrators

Dishes are great for concentrating dispersed energy to a focal point. Take a look at any residential neighborhood, and you are bound to see a menagerie of dishes (the state flower of Virginia) sticking out of the side of houses everywhere! What do you think these dishes are doing? Acting as concentrators! They take the waves emitted from satellites far above the earth's surface, and concentrate them into a focal point which strengthens the signal. Similarly, you might have seen some of the world's great radio telescopes perched up high upon hillsides. These are doing exactly the same thing, taking the signal from a wide area, and focusing it down to a small point. They are "concentrating" the weak signals from outer space to a fine point where they can be processed.

Solar concentrators using parabolic dishes are exactly the same, the difference being the medium used to coat the dishes. Rather than being reflective to radio waves, the coatings used to coat a parabolic solar reflector are mirrors.

Again this idea is not particularly new, in fact, back in the 1800s a Frenchman by the name of Augustin Mouchot was actively experimenting with using solar dishes to concentrate the sun's energy. Mouchot was concerned that coal was all going to be used up and that "Peak Coal" was approaching. He said at that time "Eventually industry will no longer find in Europe the resources to satisfy its prodigious expansion . . . Coal will undoubtedly be used up." One of Mouchot's solar concentrators can be seen in Figure 8-10. A little later in 1882, Abel Pifre, Mouchot's assistant, demonstrated a printing press in the Tuilleries Garden, Paris, powered by the sun, using a 3.5 m diameter concave concentrating dish. At the focus of this concentrating furnace,

Figure 8-10   *One of Mouchot's solar furnace dishes.*

was a steam boiler which provided steam for the printing press. A woodcut drawing of this press is shown in Figure 8-11.

Dishes are great for concentrating dispersed energy to a focal point (Figure 8-12a and b). Take a look at Figure 8-13 which shows parallel rays of light, entering a parabola and being focused to a point.

Note

If you want a cheap source for a solar parabolic mirror, the University of Oxford produce a solar energy kit (Figure 8-14), which is inexpensive, and comes with a budget plastic parabolic mirror.

Figure 8-11   *Pifre's solar printing press.*

Figure 8-12a and b *Parabolic mirrors take incoming parallel light (from the sun) and focus it to a point.*

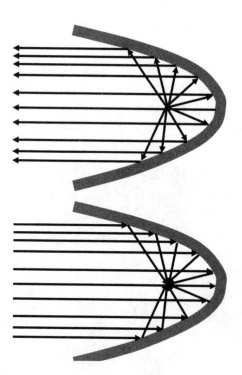

Figure 8-13 *Diagram showing how parabolas focus light to a point.*

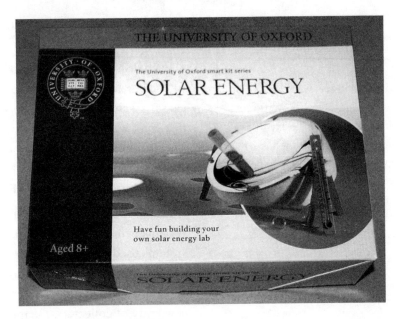

Figure 8-14   *University of Oxford solar energy kit.*

Project 17: Build Your Own Parabolic Dish Concentrator

You will need

- Old satellite dish
- Bathroom/kitchen tile adhesive
- Small mirror tiles

Tools

- Adhesive comb
- Spreader

> ### Note
>
> You need to purchase some tile adhesive—the sort of stuff you would use when applying ceramic tiles onto your walls at home.
>
> You will need to choose a tile adhesive which is waterproof, as a non-waterproof tile adhesive will not stand outside use—for this purpose kitchen/bathroom adhesive is strongly recommended.

This is an incredibly easy way to make a parabolic dish concentrator, and even better, it recycles old stuff! Take a satellite dish, and dunk your adhesive comb into the bathroom/kitchen tile adhesive. Working from the center of the dish outwards, spread the adhesive using the "comb" side of the spreader.

What the comb does is apply the adhesive in a ridged manner, this means that when you press the tiles into the adhesive, they have room to settle and even themselves out. If you just apply straight flat adhesive, when you try to push the tiles in, adhesive will ooze out everywhere and make a mess. As you work from the center, keep adding more tiles, trying as best you can to keep them in line with the plane of the parabolic satellite dish.

> ### Caution
>
> I strongly recommend that you perform this operation in your garage or in a shaded area, as with the addition of more mirrors and a little sunlight, a focal point can quickly develop which has the potential to burn you while you are working!

Note

If you are messy and get adhesive everywhere, you want to wait a little while until the tiles are firmly in place, but not so long as for the adhesive to dry, as it will only be harder to get off once it has set. To remove adhesive while it is still wet, you need a moist cloth, which you can wipe over the surface of your mirror tiles, taking off any excess adhesive with the cloth.

Free energy?

Solar dish collectors take the immense power of the sun, over the area of a dish, and concentrate that energy by means of reflectors to a central point.

At the end of 2004, Sandia National Laboratory announced that they were working with Stirling Energy Systems to build and test a six-dish array. These six dishes would be capable of producing 150 kW of power during the day, enough to power 40 homes.

Each dish comprises 82 individual mirrors all focused to a single central point (Figure 8-15). This causes a massive amount of heat to be generated at that point which is used to drive a Stirling engine. The Stirling engine produces mechanical movement, which is converted to electrical energy by a conventional generator arrangement (Figure 8-16).

One of the problems inherent with solar dish systems is that they must track the sun—older systems used really heavy mirrors which meant that the motors required to track the sun had to be big and beefy and drew a lot of energy. With this new array of collectors, the mirrors have been designed with a honeycomb structure so they are strong, and yet very light indeed.

This is said to be the largest array of solar dishes in the world, but big plans are afoot. Eventually, when the technology is fully proven, massive arrays of 20,000 units are imagined filling vast fields and plains—producing free energy from the sun (Figure 8-17).

Figure 8-15 *Solar dish engine system under test.* Image courtesy Sandia National Laboratories/Randy Montoya

Figure 8-16 *10 kW solar dish Stirling engine water pump.* Image courtesy Sandia National Laboratories/ Randy Montoya.

Figure 8-17 *Artist's rendering of a field of solar engines.* Image courtesy Sandia National Laboratories/ Randy Montoya.

Project 18: Experiment with Fresnel Lens Concentrators

You will need

- Fresnel lens
- Feather
- Small piece of rubber
- Wax candle
- Photovoltaic cell
- Multimeter

Imagine trying to build a large lens to cover a meter square in order to concentrate the sun. What would you build it out of? Well for a start, the lens would be physically quite big if it had to cover a meter square, it would also use quite a large volume of material. This is not a particularly efficient way of doing things. Far better to build a lens that uses less material. This has a number of advantages. First of all, it uses less material. As a result of this, the lens is not only cheaper, but also lighter. This means if our lens is actively tracking the position of the sun by a mechanism, the mechanism can be lighter duty, as it does not need to move such a heavy load.

Note

Very large Fresnel lenses can generate *truly awesome* power—start the experiments in this chapter with smaller Fresnel lenses such as those used as "magnifying bookmarks" before graduating to larger lenses!

Where can I get a Fresnel lens?

Here are a few ideas for procuring a Fresnel lens, both second hand and new, cost varies widely:

Car reversing lenses are a great source, like *those* pictured in Figure 8-18, these are often fairly small with a fairly coarse lens, but will certainly do the job and provide many fun hours of experimentation!

Fresnel lenses are often sold in small credit card or slightly larger sized flat plastic printed versions in bookshops. They are often sold as a bookmark, which doubles up as a text magnifier for those with poor sight. These lenses are not normally that large; however, they have quite a finely ruled lens structure.

Overhead projectors are another great source of Fresnel lenses. If you can find an old projector which is being discarded, the Fresnel lens is the surface on which you would place the transparency. As many people are now switching to video projectors and presentation software, colleges and schools are often great places to find unwanted overhead projectors.

Old large screen projection televisions are another item that sometimes use large Fresnel lenses to

Figure 8-18 *Car reversing Fresnel lens.*

make the picture larger. This will involve a bit of disassembling—so make sure you are with someone who knows what they are doing. Try to find an old broken set, not your father's latest HDTV wonder if you want to live to see your next birthday!

Also, for a nice-sized meaty Fresnel lens, you can often find plastic screens that you put in front of your TV in order to make it appear bigger.

If all else fails, a quick Google search will throw up a few results for optical suppliers. There are a lot of vendors selling kits to make large-screen projector TVs from an old screen—these lenses are often very overpriced. Online auction sites are another good source, or school science catalogs. There are also a couple of entries in the Supplier's Index (see Appendix) for new Fresnel lenses.

How does a Fresnel lens work?

To understand how a Fresnel lens is constructed and works, we are going to need to do a little thought experiment. Picture this. You have a glass lens which is flat on one side, and round on the other. We are now going to use a tool to remove material from the center point of the lens. The tool has a flat end. We are going to remove material until the corners of the flat-ended tool just begin to penetrate the round surface. We are now going to use another larger tool to remove material from a circle around the last. We are going to do this until the tool just starts to break through the surface. We are going to keep doing this with progressively larger tools until we are left with a hollowed lens.

If you were to look at the inside of this lens, what you would see is a series of flat "steps" cut in concentric circles. Now imagine flattening out these concentric circles so that they all lay in the same plane. What you have constructed in your mind is a Fresnel lens.

Take a look at Figure 8-19, it shows how a Fresnel lens is simply a normal lens with the unnecessary glass removed and flattened. It is important to note, that although Fresnel lenses tend

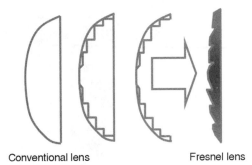

Conventional lens
Fresnel lens

Figure 8-19 *Diagram showing how a Fresnel lens compares to a conventional lens.*

to be lighter, they do not possess the same optical clarity as ordinary lenses—which is why they are not used in cameras or microscopes.

Take a look at Figure 8-20. It demonstrates that although our Fresnel lens is only a thin sheet of plastic, it can magnify things significantly.

Now try and use your Fresnel lens as a solar concentrator, hold it above a piece of paper until you form a bright white dot of sunlight (Figure 8-21). Notice how much brighter the concentrated dot is compared with the rest of the paper, which is simply illuminated by the sun.

Figure 8-20 *A thin Fresnel lens shown magnifying.*

Figure 8-21 *Concentrating solar energy using a Fresnel lens.*

A few experiments that you can do with a Fresnel solar concentrator

Use a thermometer to measure the temperature of the point where the sun's energy is concentrated—see what difference it makes if the bulb is covered in tin foil or black paper.

The intensity of the concentrated light might be enough to singe a feather, or even a thin shaving of rubber from a balloon or latex glove.

Try shining the beam of light onto a photovoltaic cell connected to a multimeter and load—see how it affects the amount of power produced.

You might also want to see if you can melt a wax candle using the power of your concentrated light.

Chapter 9

Solar Pumping

Pumping water is an essential task—we need water to drink, wash, cook, and sanitize and irrigate with. Water can be used for utility, or it can be used for dramatic effect, creating tranquility and pleasantness in our surroundings.

Using solar energy to pump water makes quite a bit of sense. Our demand for water often rises when the sun is shining. Think of agriculture—there is more sun in the summer, and that is when we want pumped water to irrigate our crops.

We can use water features to enhance our environment, water naturally has a calming destressing effect, and its importance is emphasized by disciplines such as Feng Shui.

Water can be used to add prestige to an area. The U.K. Centre for Engineering and Manufacturing Excellence (CEME) has a fountain outside, powered by solar photovoltaics on the roof. This can be seen in Figure 9-1.

Similarly at home, you use your water features in the garden when the sun is shining, not when the sky is gloomy and the weather overcast.

In this sort of application, the intermittency of solar energy does not matter so much.

Also, water can be stored relatively easy. When we actually pump it to our location doesn't really matter, as it can happily sit there in a tank. This means that we can use a supply tank to even out some of the intermittency problems.

There are other solutions to the problem; even in low light, we can harness the energy that the sun produces and store it in capacitors. When the energy stored builds up to a sufficient level, a small amount of pumping can be performed and the cycle repeats again. This is shown in the display at the Centre for Alternative Technology, U.K. (Figure 9-2).

This has some interesting consequences for our energy supply. The pumped power station at Dinorwig, Wales, draws water up into a large reservoir using excess power from the grid. When there is a shortage of power, that water is allowed to flow down hill through hydroelectric generators, producing power as it does so.

As we can see, there are many cogent reasons for using solar energy to pump our water—now let's move on to some practical projects:

Figure 9-1 *Photovoltaic-powered fountains enhance the CEME, U.K.*

Figure 9-2 *The solar pumping display at the Centre for Alternative Technology, U.K.*

Project 19: Build a Solar-Powered Fountain

Adding a solar-powered fountain to your school, home, or office is a great way to create a peaceful relaxing atmosphere. The modern world demands a lot of us, and it is nice to have some space where we can quietly go and relax and listen to the soothing sound of trickling water.

First of all, you need to decide what sort of fountain you want. There are a number of different options here—you might want a plume of water jetting into the air, you might like to add a trickling brook waterfall to your garden, or you might want a bell shaped fountain.

Once you have decided on your feature, nip down to the garden center to see what fixtures and fittings they have in store. Regular pump fittings will often specify the flow rate of the fixture, this essentially means what volume of water can be pumped at what rate, and also how high the water can be pumped. We call this the "head" of water. We will size our pump appropriately to produce this head of water.

As a rough guide, a calm, trickling waterfall will demand between 1 and 2 gallons per minute (gpm), or between 3 and 8 liters per minute (lpm) if you are working in metric.

For something with a little more razzmatazz, you might like a plume of water shooting into the air. This will generally demand a little more water, say 4–7 gpm or 15–27 lpm.

If you want the whole shebang with a cascading waterfall with a heavier current, then you really need to be considering flow rates of around 7–16 gpm, which works out as around 27–60 lpm.

Tip

If you need to convert flow of water between metric and imperial, I suggest you nip along to:
www.deltainstrumentation.com/calcs.html

Manufacturer's figures can often be optimistic and sometimes unreliable. While the minutiae sometimes don't matter, if you want to be sure and test flow rate, all you need is a gallon bucket and a stopwatch. Time how long it takes to fill up the bucket.

When you go to choose your pump, you need to realize that there are pumps, and there are pumps! The type that you require is a "DC submersible pump." A submersible pump is already water-proofed and will happily sit in the sump of your water feature. It sucks the water from the sump, and forces it out through a pipe. One of the beauties of this type of pump is that it does not require "priming," a procedure which is tiresome and often required by some other pumps.

You need to pick a pump that has a similar power rating to your solar array. Keep everything to 12 V, and if at all possible, oversize the solar panel slightly to give you adequate performance in poorer weather.

If you can't find a source of low-voltage DC pumps, then take a trip to your local chandler or boat shop. They will often sell low-voltage pumps that are used to pump water out from the bottom of boats. These are known as "bilge pumps" and shouldn't cost a lot of money.

In order to keep things nice and simple, we are just going to connect our solar panel directly to our pump (see Figure 9-5). This is nice because it allows you to visually observe the relationship between the water flowing through your feature, and the amount of sunlight falling on your panel. It does, however, mean that in overcast weather your feature will

Figure 9-3 *Waterproof solar panel showcase.*

Figure 9-4 *Solar panel on display.*

perform poorly, if at all; but then who wants to be outside when it is overcast!

Next, before committing to a feature, take your pump, dump it in a bucket of water and connect it to your panel to check that everything is working. Check this setup in good light to ensure that it is your setup, not the sun which is the problem.

Now you need to build a sump of some sort for your pump to sit in. Again, a trip to the garden center may yield a nice large-sized waterproof container, butt, or bucket.

If you are feeling particularly energetic, you could dig a hole in the ground, line it with fine sand, ensuring that there are no sharp protruding edges, and then line it with a waterproof liner.

You want your sump to be able to hold a fair quantity of water—the water in our feature will be recirculated, rather than constantly replenished. Ensure that when your submersible pump sits in the sump it is fully immersed in water.

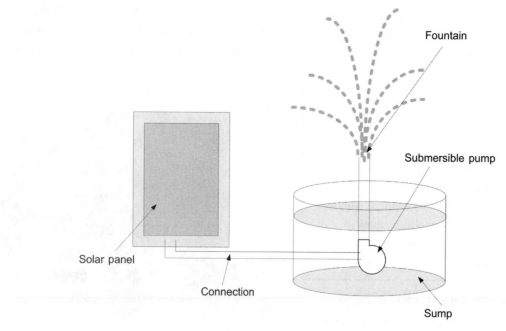

Figure 9-5 *Diagram of the solar water feature.*

One you have satisfied yourself that your pump and module work together satisfactorily, you will need to install your feature.

Things to consider

Flexible plastic tubing and jubilee clips are infinitely easier to work with than copper pipe and solder.

You will want to provide some sort of mechanical protection for your cable to ensure that it does not become chafed, or cannot easily be damaged by gardening activities such as digging. Encase the pipe in a hard plastic pipe, or mount it above ground where it can clearly be seen.

Chapter 10

Solar Photovoltaics

A solar photovoltaic device is one which takes light from the sun and turns it into electricity. In doing so, it produces no emissions or harmful waste, and does so completely silently!

The origin of photovoltaic solar cells

None of this would be possible if it hadn't been for the work of French Physicist Edmund Becquerel, who in 1839, discovered the photovoltaic effect. In fact, Becquerel is a bit of an inspiration for young Evil Geniuses wanting to experiment with solar energy, as he made his discoveries when he was only 19! In 1883, Charles Fritts, an American inventor, devised the first practical solar cell, when he took some selenium and covered it with a fine coating of gold. His cell wasn't particularly efficient, with 1% or so conversion efficiency from light to electricity; however, his design of cell later found applications as a sensor in early cameras to detect the light level—being used to "sense" light rather than to generate power in any real quantity. Albert Einstein went on to further develop the theory of the nature of light and the mechanism through which the photoelectric effect works, the discovery was considered so important that he won the Nobel Prize in 1905. Because of their high cost and low efficiency at that time, there were a lack of applications for photovoltaic cells. It was not until Bell Laboratories started looking at the idea again in the 1930s that Russell Ohl discovered the silicon photovoltaic cell. This device was patented as Patent no: US2402662 "Light sensitive device." Now the efficiency of solar cells began to increase.

The first generation of practical solar cells was horrendously expensive, and this severely limited their range of applications. The advantages of using photovoltaic cells to turn sunlight into electrical power were initially appreciated for powering satellites and space-missions. With the space race of the 1950s and 1960s, there was suddenly a good application for solar cells—despite the cost, solar cells were suitable for generating energy in the remote reaches of space. Vanguard 1 was launched on March 17, 1958, and was the first artificial satellite to employ solar photovoltaic cells. With the injection of funding and research that came with the space race, solar cells began to come into their own. Over the years that followed, solar photovoltaic technologies have been refined and developed and new techniques explored. We are now at the point where we have a range of different photovoltaic technologies, and we will explore these now.

Solar cell technologies

There are a number of different technologies that can be used to produce devices which convert light into electricity, and we are going to explore these in turn. There is always a balance to be struck between how well something works, and how much it costs to produce, and the same can be said for solar energy.

We take solar cells, and we combine them into larger units called "modules," these modules can again be connected together to form arrays. Thus we can see that there is a hierarchy, where the solar cell is the smallest part (see Figure 10-1).

Figure 10-1   *Cells, modules and arrays.*

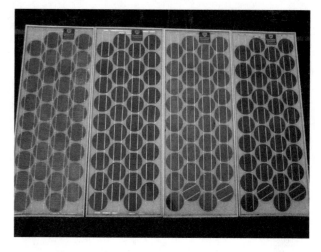

Figure 10-2   *Monocrystalline solar cells made into a panel.*

In this chapter we are going to look at the structure and properties of solar "cells," but bear in mind, when combined into modules and arrays, the solar "cells" here are mechanically supported by other materials—aluminum, glass, and plastic.

One of the materials that solar cells can be made from is silicon—this is the material that you find inside integrated circuits and transistors. There are good reasons for using silicon, it is the next most abundant element on earth after oxygen. When you consider that sand is silicon dioxide (SiO_2), you realize that there is a lot of it out there!

Silicon can be used in several different ways to produce photovoltaic cells. The most efficient solar technology is that of "monocrystalline solar cells," these are slices of silicon taken from a single, large silicon crystal. As it is a single crystal it has a very regular structure and no boundaries between crystal grains and so it performs very well. You can generally identify a monocrystalline solar cell, as it

appears to be round or a square with rounded corners; you can see monocrystalline solar cells in Figure 10-2.

One of the caveats with this type of method, as you will see later, is that when a silicon crystal is "grown," it produces a round cross-section solar cell, which does not fit well with making solar panels, as round cells are hard to arrange efficiently. The next type of solar cell we will be looking at, also made from silicon, is slightly different, it is a "polycrystalline" solar cell. Polycrystalline cells are still made from solid silicon; however, the process used to produce the silicon from which the cells are cut is slightly different. This results in "square" solar cells. However, there are many "crystals" in a polycrystalline cell, so they perform slightly less efficiently, although they are cheaper to produce with less wastage.

Now, the problem with silicon solar cells, as we will see in the next experiment, is that they are all effectively "batch produced," which means they are produced in small quantities, and are fairly expensive to manufacture. Also, as all of these cells are formed from "slices" of silicon, they use quite a lot of material, which means they are quite expensive.

Now, there is another type of solar cells, so-called "thin-film" solar cells. The difference between these and crystalline cells is that rather than using crystalline silicon, these use chemical compounds to semiconduct. The chemical compounds are deposited on top of a "substrate," that is to say a base for the solar cell. There are some formulations that do not require silicon at all, such as CIS (copper indium diselenide) and cadmium telluride. However, there is also a process called "amorphous silicon," where silicon is deposited on a substrate, although not in a uniform crystal structure, but as a thin film. In addition, rather than being slow to produce, thin-film solar cells can be produced using a continuous process, which makes them much cheaper.

However, the disadvantage is that while they are cheaper, thin-film solar cells are less efficient than their crystalline counterparts. Some different solar photovoltaic technologies are compared in Table 10-1. Figures are given for the efficiency of the cell technology, and the average area of cells required to generate 1 kW peak power when facing in the right direction!

When looking at the merits of crystalline cells and thin-film cells, we can see that crystalline cells produce the most power for a given area. However, the problem with them is that they are expensive to produce and quite inflexible (as you are limited to constructing panels from standard cell sizes and cannot change or vary their shape).

Table 10-1
Efficiency of different cell types

Cell material	Efficiency	Area required to generate 1 kW peak power
Monocrystalline silicon	15–18%	7–9 m²
Polycrystalline silicon	13–16%	8–11 m²
Thin-film copper indium diselenide (CIS)	7.5–9.5%	11–13 m²
Cadmium telluride	6–9%	14–18 m²
Amorphous silicon	5–8%	16–20 m²

Source data: Deutsche Gesellschaft fur Sonnenenergie e.V.

By contrast, thin-film cells are cheap to produce, and the only factor limiting their shape is the substrate they are mounted on. This means that you can create large cells, and cells of different shapes and sizes, all of which can be useful in certain applications.

We are now going to take a detailed look at making two different types of solar cell, one will be a crystalline solar cell, and the other a thin-film solar cell. Both of the experiments are designed to be "illustrative," rather than to actually make a cell with a useful efficiency. The technology required to make silicon solar cells is out of the reach of the home experimenter, so we are going to "illustrate" the process of how a solar cell is made, using things you can find in your kitchen. For thin-film solar cells, we are going to make an *actual* solar cell, which responds to light with changing electrical properties; however, the efficiency of our cell will be very poor, and it will not be able to generate a useful amount of electricity.

How are crystalline photovoltaic cells made?

In this section we are going to look at how photovoltaic cells (PV) are made. However, rather than taking a dull, textbook approach, we are going to make the whole process fun by doing some practical kitchen experiments that mimic the process that happens in solar cell factories all around the world.

How do they work?

First of all, let's cover a little bit of the theory.

Ordinary silicon forms into a regular crystalline structure. If you look at Figure 10-3, you can see the way that the silicon atoms align themselves into a regular array.

To make silicon "semiconducting," we can take a little bit of another chemical, in this case boron, and

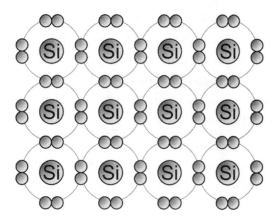

Figure 10-3 *Plain old silicon—its atomic structure.*

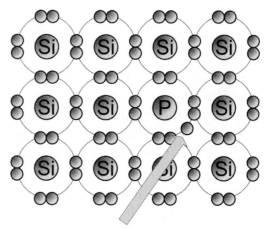

Extra electron

Figure 10-5 *Silicon doped with phosphorus—note the spare electron.*

introduce it to the silicon. Where there is a boron atom, there is also a missing electron. This creates a "hole" in the outer shell of the boron atoms and its neighboring silicon atom (Figure 10-4).

If we add a little bit of phosphorus to our silicon, we get the opposite effect, a "spare" electron (Figure 10-5), which doesn't quite know where to fit in. As a result, it sort of "lingers uncomfortably" waiting for something to happen.

Now, we can use these two types of "doped" silicon to make semiconducting devices, in this case "photovoltaic cells."

A photovoltaic solar cell is a bit like a sandwich. It is made from layers of different types of silicon, as illustrated in Figure 10-6.

Starting from the base, we have a large contact. Then on top of this we have a layer of p-type silicon, a junction called the space charge region

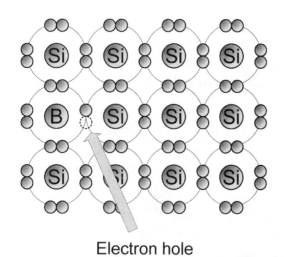

Electron hole

Figure 10-4 *Silicon doped with boron—note the missing electron.*

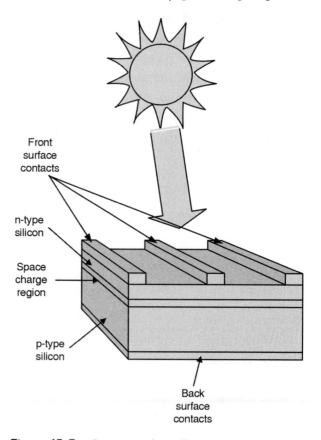

Figure 10-6 *Cutaway solar cell.*

where the magic occurs, and a slice of n-type silicon on top.

On top of all this is layered a grid electrode, which does the job of making the other contact.

Now, photons from the sun hit our solar cell, and in doing so "spare" negatively charged electrons,

are "knocked" across the boundary between p- and n-silicon, which causes a flow of electrons around the circuit.

We are now going to look at how the silicon for these solar cells is manufactured, using some things you can do at home.

Project 20: Grow Your Own "Silicon" Crystals

You will need

- Plastic coffee jar (empty)
- Skewer
- Hardboiled egg
- Sugar
- Food coloring

Tools

- Compass
- Egg slicer

To make a photovoltaic cell we need silicon, this project is going to show you how solar cells are produced from crystalline silicon. The words "crystalline silicon" should indicate to you that this type of solar cell is made from crystals of silicon. We saw earlier how silicon aligns itself into a regular crystalline array, now we are going to look at growing this crystal.

In industry, silicon crystals are grown to form a uniform cylinder of silicon which is used as the base material for crystalline solar cells. There is plenty of silicon about on the earth, in fact, as mentioned previously, after oxygen it is the second most abundant element. When you think that sand and quartz all contain silicon and then imagine the amount of sand in the world, you begin to realize

that we are not going to run out of silicon in a hurry!

The problem with sand is that it also contains oxygen in the form of silicon dioxide, which must be removed.

The industrial process used to produce silicon requires temperatures of around 3270°F (which is about 1800°C). Obviously we can't experiment with these sorts of temperatures at home—but we can recreate the process!

If you don't want to get the individual bits and bobs, a couple of educational scientific vendors sell rock-growing kits. These links are to suppliers of kits of parts:

- scientificsonline.com/product.asp?pn=3039234& bhcd2=1151614245
- www.sciencekit.com/category.asp_Q_c_E_737919
- www.scienceartandmore.com/browseproducts/ Rock-Candy-Growing-Experiment-kit.html

If you want to do it all yourself, then you can see from Figure 10-7 that the process is a relatively easy one! You are going to need a saturated sugar solution, this will sit in the lid of your coffee jar. Now, take a large crystal of sugar, often sold as "rock sugar" and "glue" it to the end of the skewer. Next, drill a hole the same diameter as the skewer, and poke the skewer through the bottom of the coffee jar. Stand it on a windowsill and lower the crystal into the saturated sugar solution. Over some

Figure 10-7   *Growing sugar crystals.*

time, crystals should start to grow—pull the skewer up slowly, bit by bit, so that the growing crystal is still in contact with the sugar solution.

This is just like the way that silicon is grown. The silicon is drawn up slowly from a bath of molten hot silicon (which is analogous to our saturated sugar solution). This is shown in Figure 10-8.

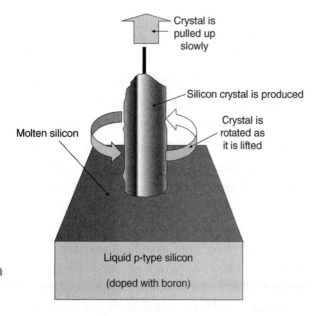

Crystal is pulled up slowly

Silicon crystal is produced

Crystal is rotated as it is lifted

Molten silicon

Liquid p-type silicon

(doped with boron)

Figure 10-8   *Growing silicon crystals.*

Figure 10-9   *Slicing eggs.*

Once this large crystal of silicon has been manufactured, it must be cut into slices to manufacture the solar cells. I like to think of this a bit like the way an egg is sliced to make sandwiches by an egg slicer—see the analogy in Figures 10-9 and 10-10.

"Slice and dope" your silicon crystals

Slicing an egg with an egg slicer is much like the process that happens when a solar cell is manufactured. Each slice of silicon is then called a "wafer."

We now need to create a p–n junction in the wafer; to do this phosphorus is diffused into the surface of the silicon. Dip your egg into some food coloring or beetroot juice, and you will see that the juice covers one surface of the egg slice. Now, imagine that slice of egg were a solar cell, with the beetroot-soaked face pointing toward the light. Imagine an electrical contact on either side of the egg slice is connected to our circuit. The photons

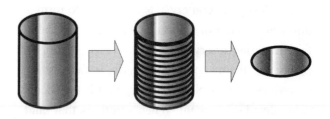

Figure 10-10   *Slicing silicon.*

hit the colored side, which is "doped" with phosphorus to produce some extra electrons. By giving these electrons additional energy from the photons, they are able to "jump" the gap, across to the "boron" doped silicon (the plain old egg) where they fill the "holes" where there are electrons missing from the atomic structure. With a steady stream of photons, hitting the cell, a heavy stream of electrons are encouraged to migrate across the p–n junction, then travel around the circuit doing useful work!

Now these cells can be integrated into larger modules, or even arrays, to produce more power.

Figure 10-11 *Doping with phosphorus.*

Now we have looked at the technology of crystalline solar cells using silicon, let us turn our attention to thin-film solar cells.

Project 21: Build Your Own "Thin-Film" Solar Cell

Note

First of all a little disclaimer ... the solar cell you are about to build here is *horribly, horribly* inefficient. Please do not have any plans to use these to power your home. The amount of current that they produce is *very small* and not economically exploitable. While this is a shame, this project is very interesting, educational and helps you get to grips with the photoelectric effect.

You will need

- Copper sheeting
- Clear Plexiglas/Perspex/acrylic sheeting
- Some thin wood strip
- Copper wire
- Duct tape

Tools

- Metal guillotine (optional)
- Bandsaw (optional)
- Tin snips
- Electric ring hob

First of all, cut a square of the copper sheeting so that it is about 6–8 in. square in size. It is much easier to do this with a metal guillotine (Figure 10-12); however, if you haven't got access to this sort of equipment, tin snips will work just fine.

When you have done this, wash your hands thoroughly and dry them. You need to remove any grease or oil from your hands that could cause problems with the next step of the process. Remove any grease or detritus from the copper sheeting. Next, take a piece of emery cloth (see Figure 10-13), and thoroughly sand down the piece of copper on both sides to remove the top layer of oxidized copper.

Figure 10-12 *Cutting the copper with a guillotine.*

Figure 10-13 *Cleaning the copper with emery cloth.*

This will leave you with nice bright shiny red copper underneath.

You now need to heat treat the copper, in order to form an oxide coating on top. It may sound counterintuitive that we have just removed all the oxide and now we are going to put oxide back on, but the oxide coating we will be applying will be a film of "cuprous oxide."

You will need an electric hob to do this. If you have any "heat proof gloves" and metal tongs, this might be the time to get them in order to handle the metal while hot.

You need to turn the burner to the highest setting, with the sheet of copper just placed on top. Observe the changes to the copper carefully, they are *very* interesting.

As you heat the copper, it takes on a lovely vivid patina of different colors. Obviously, the pages here are black and white, so I can't show you, but if you look at Figure 10-14 a–e you will see the changes that the plate goes through.

Figure 10-14 *The shiny copper plate on the burner.*

Hint

If you have access to nitric acid, you can use this as a superior method for removing the upper cupric oxide layer.

You will see a black crusty oxide form on top of the copper plate. If you leave the plate to cool slowly, the crusty layer should become fairly fragile and separate easily from the underlying copper. When you have allowed the plate to cool thoroughly, give the plate a firm bang edge-on to a hard surface. Some of the oxide will pop off. Rub the oxide *gently* with your fingers under a tap, and you will find most of the black layer of oxide comes off easily. If any bits are stubborn, *do not under any circumstances scour them*, as we do not want to damage the fragile surface.

Under this black layer of oxide, you will find another layer of a reddish orange rust color. This is the layer which is "photosensitive" and will make our thin-film solar cell work.

Make a spacer now from some thin strips of wood (Figure 10-15). I used duct tape to join my pieces of wood together—do not use metal fixings as they could react electrolytically with the other components of the cell.

We are now going to make another electrode. It has to have the property that it does not touch the

Figure 10-15 *The spacer piece.*

Figure 10-16 *Perspex and duct tape.*

other piece of the solar cell, and allows light to hit the surface. We are going to use salt water as our other electrode, making contact with the whole surface of the thin film cell, yet conducting electricity. We are then going to immerse another copper wire to make the connection. You could equally use another piece of copper plate around the outside of the thin-film cell, but not touching our oxidized copper.

In a commercial thin-film cell, tin oxide is commonly used as the other electrode, as it is clear and yet conducts electricity.

Now take a piece of Perspex to act as a cover plate, and stick a strip of duct tape on either side, as shown in Figure 10-16.

We are going to stick our other electrode wire to this piece of Perspex.

In Figure 10-17, I have used thickish wire for clarity, with few actual zigzags so that you can clearly see what is going on. To optimize the performance of your solar cell, you want to make the conductor large. To this end, you are better using *lots* of thinner gauge wire in a much finer zigzag pattern—this will still allow the light to get through, but at the same time gives a large conductor area.

You can experiment with different types of wire and copper—the trick is to try and maximize the surface area of the copper, while trying to block as little light as possible from reaching the solar cell.

Fold the duct tape over and stick the wire to the plate.

We are now going to combine the electrode plate with the space. Again, duct tape makes this a nice easy job (Figure 10-18).

Next, we are going to take the copper plate, and stick duct tape to one side, with the sticky side of the tape facing the same direction as the layer of red copper oxide (Figure 10-19).

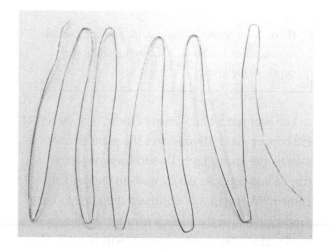

Figure 10-17 *Wire electrode.*

Figure 10-18   *Perspex plate and electrode combined.*

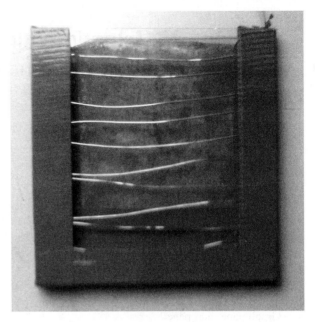

Figure 10-20   *The finished solar cell.*

Combine the plate and the front module to make the finished solar cell (Figure 10-20).

Now, take a little salt water, and fill the void between the Perspex front section and the copper plate. Seal the module with duct tape all round to prevent leakage.

Finally, connect your module to a multimeter, find a bright light source, and explore some of the electrical properties of your solar cell.

Figure 10-19   *The copper plate with duct tape fixings.*

Chemical data file: cuprous oxide

Cuprous oxide (red)

Formula	Cu_2O
Molecular weight	143.08
Physical appearance	Red to reddish brown powder

Experiments with photovoltaic cells

In this project we will be performing a range of experiments with photovoltaic cells that allow us to learn something about their characteristics and how they perform in different applications.

The experiments in this project could form a great basis of a science fair stand or poster display.

Project 22: Experimenting with the Current–Voltage Characteristics of a Solar Cell

You will need

- Light source
- Photovoltaic cell
- Voltmeter
- Ammeter
- Variable resistance
- Graph paper and pencil

  or

- Computer with spreadsheet package

We can learn a lot about solar cells' electrical characteristics by plotting the "current–voltage" curve of the device.

To carry out the experiment, we will need to ensure that the solar cell receives constant illumination all the time. Use a bright lamp, and position it a fixed distance above the solar cell.

Set up the circuit as shown in Figure 10-21.

We are now going to adjust the variable resistor from one extreme to the other, noting how the readings on the voltmeter and ammeter change as we do so. At this point you need to make careful notes as to the current and the voltage at each stage. You can do this on paper, or, if you have a PC handy, on a spreadsheet. Try and take at least 15 or so different readings to help you plot an accurate curve.

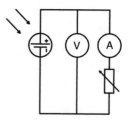

Figure 10-21 *Circuit to determine the current–voltage curve of a single solar cell.*

Now plot the points on your graph paper, or by using the chart wizard on a spreadsheet program. Compare your graph to Figure 10-22. The graph tells us how the solar cell will perform when different loads are applied.

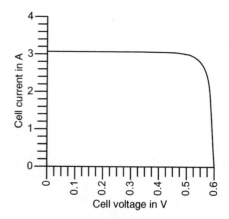

Figure 10-22 *Current–voltage characteristics of a single solar cell.*

Project 23: Experimenting with Current–Voltage Characteristics of Solar Cells in Series

You will need

- Light source
- Three photovoltaic cells
- Voltmeter
- Ammeter
- Variable resistance
- Graph paper and pencil

  or

- Computer with spreadsheet package

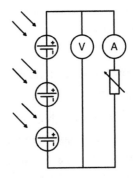

Figure 10-23 *Circuit to determine the current–voltage curve of solar cells in series.*

Now we are going to repeat the experiment above, but we are going to do it three times.

Set up the circuit as shown in Figure 10-23. First using one cell, then two, then three.

You can reuse your result for above for the single solar cell, but we are now going to add two additional lines to our graph—one for two solar cells connected in series, and another for three solar cells in series.

What can we see from the results (Figure 10-24)? Well, it is clear that when we add multiple solar cells in series, the voltages "add up." However, the current produced remains the same.

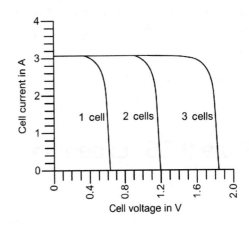

Figure 10-24 *Current–voltage curve of solar cells connected in series.*

Project 24: Experimenting with Solar Cells in Parallel

You will need

- Light source
- Three photovoltaic cells
- Voltmeter
- Ammeter
- Variable resistance
- Graph paper and pencil

  or

- Computer with spreadsheet package

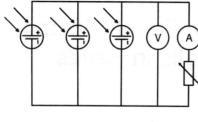

Figure 10-25 *Circuit to determine the current–voltage curve of solar cells in parallel.*

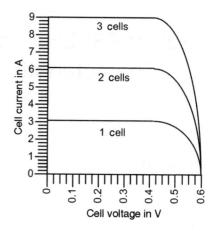

Figure 10-26 *Current–voltage curve of solar cells connected in parallel.*

We are now going to connect solar cells in parallel and repeat the experiment.

Again, we will end up with a graph with three lines. Make a prediction now! How do you expect this graph to differ from the one when we connected solar cells in series?

The solar cells will be connected in accordance with Figure 10-25. First connect one cell, then two in parallel, then three!

Now plot the graph from the points that you obtained (Figure 10-26) and compare it to Figure 10-24.

How do the two graphs differ? Well, it can be seen that in the parallel plots, the voltage remains the same throughout, and it is the current that changes—contrast this to the series experiment where it was the voltage that changed.

Project 25: Experiment with the "Inverse Square Law"

You will need

- Light source
- Photovoltaic cell
- Voltmeter
- Ammeter
- Variable resistance
- Graph paper and pencil
  or
- Computer with spreadsheet package

The inverse square law says that for each unit of distance you move a light away from a solar cell, the amount of received light is equal to the inverse of the square of that distance (Figure 10-27).

As we are trying to measure the light only from a point source, it is a good idea if you can try and do this in a darkened room.

Take a single solar cell, and connect a voltmeter and ammeter across its terminals. We are going to move the light away and measure the voltage and the current produced. Remember, it is easy to find the total "power" produced by multiplying the voltage and the current together. Compare the power generated, to the distance that the light source is from the solar cell. Plot this in a copy of Table 10-2. What do we learn about the

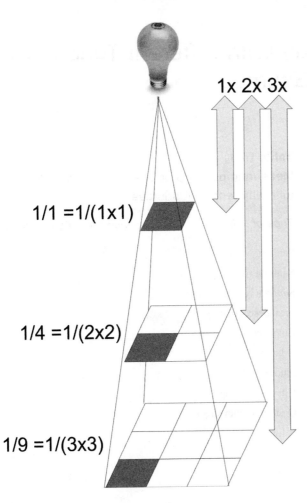

$$1/1 = 1/(1 \times 1)$$

$$1/4 = 1/(2 \times 2)$$

$$1/9 = 1/(3 \times 3)$$

Figure 10-27 *Inverse square law.*

Table 10-2

Measuring power produced by a solar cell when light is held at different distances

Distance (in.)	Distance (cm)	Load voltage (V)	Short circuit current (mA)
0	0		
2	5		
4	10		
6	15		
8	20		
10	25		
12	30		
14	35		
16	40		
18	45		
20	50		

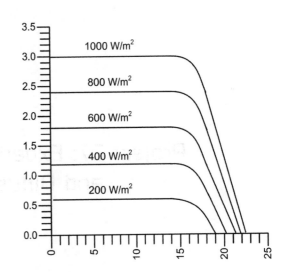

Figure 10-28 *How the current–voltage curve of the solar cell changes with varying illuminance.*

relationship between the light falling on the cell and the power generated?

Our solar cell produces more power when there is more light falling on it. We can repeat the experiment for a current–voltage curve, with different amounts of light falling on the solar cell.

What we learn is that the current–voltage curve of the cell changes depending on the amount of light falling on it. This can be seen in Figure 10-28.

Project 26: Experimenting with Different Types of Light Sources

You will need

- Light source
- Photovoltaic cell
- Voltmeter
- Ammeter
- Variable resistance

In this experiment we are going to look at the range of values for power produced from different light sources. Take your solar cell, and connect it in the same manner as when we measured current–voltage curves, and try different sources of light. Plot the results in a copy of Table 10-3. How does the power generated compare with natural light?

Table 10-3
Measuring power produced by a solar cell with different light sources

Type of light	Load voltage (V)	Current (mA)
Sunlight (sunny day)		
Sunlight (dull day)		
Sunlight (overcast day)		
Incandescent lamp		
Compact fluorescent lamp		
Fluorescent lamp		
Ultraviolet lamp		
Orange sodium street light		

Project 27: Experimenting with Direct and Diffuse Radiation

You will need

- Light source
- Photovoltaic cell
- Voltmeter
- Ammeter
- Variable resistance
- Paper to shade

The concept that we are going to get to grips with in this experiment is that reflected light can produce an awful lot of illumination and hence energy.

Look at the two types of radiation hitting the solar cell in Figure 10-29. Now, using the techniques shown in Figures 10-30 and 10-31, shade the solar cell from either direct or indirect radiation and note the amount of power that is produced.

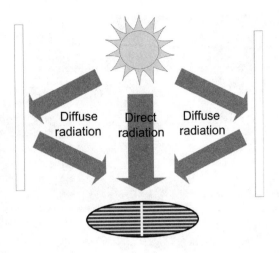

Figure 10-29   *Radiation hitting a solar cell.*

How does the amount of power produced from indirect radiation compare to that from direct radiation? There are solar cells available called "bifacial solar cells" (Figure 10-32).

These solar cells are mounted on a clear substrate to form a module. They have the advantage that they can collect light from both sides, so they can absorb direct and indirect radiation.

This means that they can absorb more power than had they just been collecting light from one direction.

In Figure 10-33, they have been mounted on the roof of a covered walkway. In this application, the solar cells are serving two purposes—generating clean energy, while keeping the rain off people walking along the walkway.

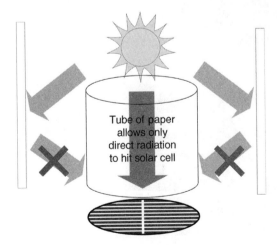

Figure 10-30   *Blocking indirect radiation.*

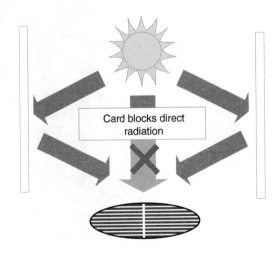

Figure 10-31   *Blocking direct radiation.*

Figure 10-32   *Bifacial solar cells.*

Figure 10-33   *Bifacial solar cells on the roof of a covered walkway.*

Project 28: Measurement of "Albedo Radiation"

You will need

- Light source
- Photovoltaic cell
- Voltmeter
- Ammeter
- Variable resistance
- Paper to shade

What is albedo radiation?

The ground is a surface just like any other, it has the capability to reflect radiation so we must not ignore it. Just think, black tarmac is bound to reflect less radiation than say a light gray concrete.

Why are we bothered—our solar cells point toward the sky don't they?

Well, yes, that is true in most cases; however, bifacial solar cells are able to accept solar radiation on both faces.

The experiment

The next experiment may seem counterintuitive, but it is very worthwhile. We are going to be measuring albedo radiation. Using your solar setup, point your PV panel at the floor and take a measurement (Figure 10-34).

What did you expect? A zero reading? In fact, as you can see, there is still a lot of energy in "indirect" radiation which is reflected from other surfaces. We saw in the last experiment how bifacial solar cells are able to collect the solar energy reflected from two faces. Therefore, in the covered walkway they can collect energy reflected from the ground (albedo) as well as from direct radiation.

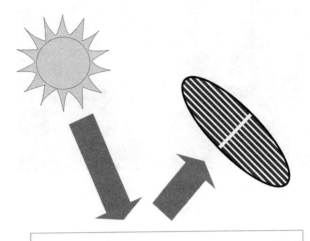

Figure 10-34 *Measuring albedo radiation.*

Applications of photovoltaic cells

We have explored some of the properties of individual photovoltaic cells and seen how light can be used to produce electricity, now let's look at some applications of solar cells.

First of all, despite electricity from solar cells currently costing much more than power from the grid, solar cells can be useful for applications where there is not a nearby electricity supply, and where a connection to the grid could potentially be quite expensive.

We saw earlier, how the first solar cells were used to power satellites in space (Figure 10-35), where other forms of power were impractical.

Figure 10-36 shows a road sign in the English countryside, the black sign above is an illuminated display, which lights up when drivers go too fast. It is lit by power produced during the day from a solar cell and also from a micro wind turbine at

Figure 10-35 *The HEESI satellite powered by solar power.* Image courtesy NASA.

Figure 10-36 *A road sign powered by renewable energy.*

day or night. The power is stored locally in batteries located in the foundation of the sign.

In addition to powering devices in remote locations with no access to the power grid, we can also construct large photovoltaic arrays, which generate a significant quantity of electricity which can be "fed into" the grid when it is not being used on-site. The great thing about photovoltaic cells, is that they can be used in place of things like roof tiles and shingles—so although we cover the building with photovoltaic cells, which are expensive, we save on the cost of the roofing material.

We can see how a solar array can be made plain and large as in this solar array at the Centre for Alternative Technology, U.K. (Figure 10-37).

Or with a little bit of thought, they can be integrated creatively into the building fabric as shown in Figure 10-38.

What does it take to solar power my home?

Producing electricity by photovoltaic cells is fairly expensive compared to other types of generation. However, when considering the "cost" of solar energy, figure in all of the carbon emissions that you aren't producing, and the toxic waste that you aren't making.

We now know that solar cells can be used to generate electricity, but the problem is getting it in a form that we can use in our homes. Sure, it is possible to run a few simple bulbs from a DC supply, but to run most of our household appliances, we need to generate electricity in a form that is suitable for them—AC.

You will notice that the output from all of our solar cells is "direct current" (see Figure 10-39). The voltage is always a fixed polarity with reference to 0 V. We can couple solar cells in series to produce a higher voltage, or in parallel to produce

Figure 10-37   *The 11 kW solar array at the Centre for Alternative Technology, U.K.*

Figure 10-38   *Photovoltaic cells creatively integrated into a building fabric.* Image courtesy Jason Hawkes.

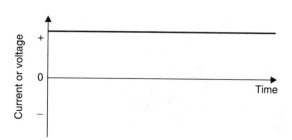

Figure 10-39 *Direct current.*

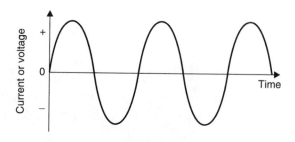

Figure 10-40 *Alternating current.*

a higher current, but we are always going to end up with DC.

By contrast, in our homes, our appliances and devices require "alternating current," AC (Figure 10-40). We see how the AC waveform differs dramatically from the steady DC line. In the United States, the frequency of this AC supply is 60 Hz, in the U.K. it is 50 Hz, it is also at a higher voltage (120 V in the U.S.A., and 230 V in the U.K.).

So, how can we take the power from our photo-voltaic cells, and turn it from "DC low voltage"

into "AC high voltage"? The answer is that we use an "inverter."

An inverter is a piece of electronics (Figure 10-41), which takes the DC supply from our solar cell and generates an AC waveform at the correct voltage and frequency for our items of mains equipment.

We need some extra devices for safety reasons, you will see in the setup that there is a mains isolator switch (as shown in Figure 10-42). This allows us to disconnect the mains from the inverter

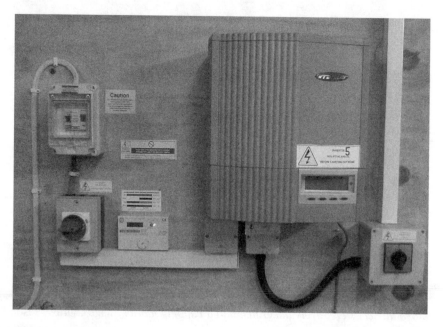

Figure 10-41 *A typical inverter setup.*

Figure 10-42 *Mains isolator switch.*

Figure 10-43 *Mains circuit breaker.*

Figure 10-44 *Solar DC isolator switch.*

in the event that we need to carry out work or maintenance.

We also need to include a mains circuit breaker to protect against overcurrents or surges, which could be potentially damaging and dangerous. A circuit breaker is shown in Figure 10-43.

And in addition, we need to be able to isolate the DC supply coming from our solar array. A DC isolator switch is shown in Figure 10-44.

It is also interesting to see how much energy our solar array is producing. This can be useful for accounting purposes, say if we are selling the solar energy back to the grid, or simply to benchmark the performance of our solar system and see if it is in line with our design predictions. A watt hour meter is shown in Figure 10-45.

Figure 10-45 *Watt hour meter.*

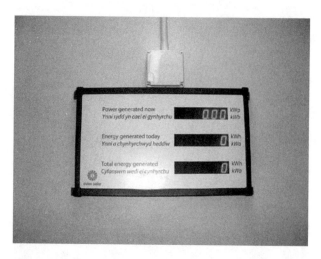

Figure 10-46 *Awelamentawe school solar display.* Image courtesy Dulas.

Of course, if we have a solar array in a public area, it is also nice to promote solar technology to others, and our solar array is a powerful tool to educate others with. At this school in Wales, in Awelamentawe (Figure 10-46), a display is prominently mounted in the main reception, to show visitors, and help educate children about, how much energy the school's solar array is producing.

Chapter 11

Photochemical Solar Cells

My sincerest thanks to Dr. Greg P. Smestad for the information and images he has provided, on which this chapter is based.

In addition to the photovoltaic solar cells that we have seen earlier in this book, there are other ways of generating electricity directly from the sun. We saw how photovoltaic solar cells rely on the photovoltaic effect that occurs at semiconductor junctions, and how the semiconductor performs the two jobs of absorbing the light and separating electrons.

One of the problems with this approach is that, because of the sensitive nature of the cells, they must be manufactured in ultra-clean conditions in order to be clean and free from defects which might impede their operation.

This works effectively; however, it is expensive.

The thing about photochemical solar cells is that they use cheap technology. Titanium dioxide is not some rare chemical that requires expensive processing, it is already produced in large quantities and used commonly; furthermore, you don't need an awful lot of it—only around 10 g per square meter. When you figure that this 10 g only costs two cents, you begin to realize that this is a solar technology with a lot of promise for the future.

In an attempt to make solar technology cheaper and more accessible, Michael Grätzel and Brian O'Regan from the Swiss Federal Institute of Technology decided to explore different approaches to the problem.

Note

The photochemical solar cell is sometimes also referred to as the "Grätzel" cell after Michael Grätzel who worked on developing the cell.

The photochemical solar cell has grown out of an expanding branch of technology—biomimickry, looking at how we can mimic natural processes to make more advanced technologies.

Rather than having a single thing to do all of the jobs, as in a conventional photovoltaic cell, photochemical solar cells mimic processes that occur in nature.

Electron transfer is the foundation for all life in cells; it occurs in the mitochondria, the powerhouses of cells which convert nutrients into energy.

Titanium dioxide, while not immediately springing to mind as a household name, is incorporated in a lot of the products that we use every day. In paints, as a pigment, it is known by its name titanium white. It is also used in products such as toothpaste and sunscreen. Titanium dioxide is great at absorbing ultraviolet light.

Tip

You might find titanium dioxide referred to as "Titania" in some references.

How do photochemical solar cells work?

Take a peek at Figure 11-1, in the top image, we can see the energy transfers taking place—the light striking our photochemical solar cell, generating energy and turning the shaft of the electric motor, which is connected to our cell. The radiated energy from the sun in the form of light, is being transformed through a chemical process into electrical energy, which travels through the circuit to the motor, where electromagnets turn the electrical energy into movement (kinetic energy).

We need to look at the cell in a little more depth to understand the chemical processes that are taking place in it in order to generate the electricity.

The dye when it is excited by light injects an electron into the titanium dioxide with which the plates are coated and semiconducts.

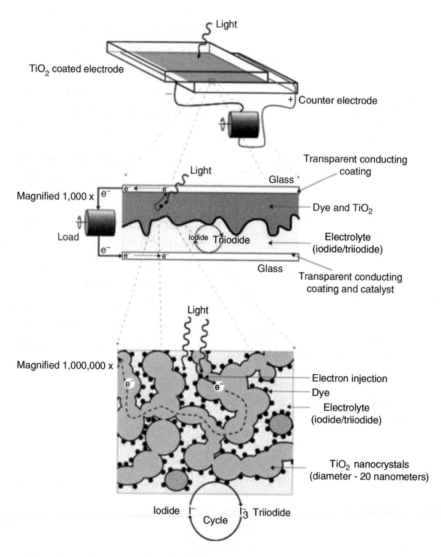

Figure 11-1 *How a photochemical solar cell works.* Image courtesy Greg P. Smestad.

Project 29: Build Your Own Photochemical Solar Cell

Online resources

Point your browser toward

www.solideas.com/solrcell/cellkit.html

for more information on dye sensitized photochemical solar cells and where you can obtain a kit of the parts featured in this project.

You will need

- Berries
- Motor
- Alligator clips
- Wires
- Nanocrystalline TiO_2 Degussa P25 powder in mortar and pestle
- Glass plates

Tools

- Petri dishes
- Tweezers
- Pipette
- Pencil

We need to get our titanium dioxide ground down so that the particles are as small as possible—this maximizes surface area, and so allows our reactions to take place quickly. To do this, we will need the mortar and pestle mentioned in our "tools" list (Figure 11-2). Be careful not to inhale any of the fine titanium dioxide powder as you are grinding, as it won't do you any good!

Now that we have prepared our suspension of titanium dioxide, we need to coat it onto our glass plate using a glass rod. This is shown in Figure 11-3.

Figure 11-2 *Grinding the nanocrystalline titanium dioxide.* Image courtesy Greg P. Smestad.

The next thing that we need to do is sinter the titanium dioxide film in order to reduce its resistivity. This is shown in Figure 11-4. To do this, we hold it in a Bunsen flame and allow the gas to do the work! We need to hold the plate at the tip of the flame where the temperature is approximately 450°C or 842°F.

Hold it steady for around 10–15 minutes.

Now you need to produce the dye which will sensitize our photochemical solar cell. There are

Figure 11-3 *Using a glass rod to spread the suspension onto the plate.* Image courtesy Greg P. Smestad.

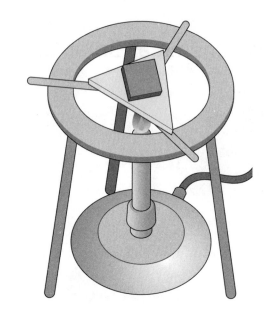

Figure 11-4   *Firing the film of titanium dioxide in order to sinter it.* Image courtesy Greg P. Smestad.

Figure 11-5   *Coating the plate in berry juice.* Image courtesy Greg P. Smestad.

a number of suggestions for different substances which can be used for this cell. You can try:

- Blackberries
- Raspberries
- Pomegranate seeds
- Red hibiscus tea in a few ml of water

To produce the dye, you need to take the substance you are going to make the dye from, and crush it in a small saucer or dish. Once this has been done and a nice fluid has been produced, take the plate which has been coated in titanium dioxide, and immerse it in the dye. The titanium dioxide film should now be stained a deep red to purple color and the color distribution should be nice and even. If this is not the case, you can immerse the plate in the dye again. Once you have finished staining the plate, take a little ethanol and wash the film and then with a tissue, blot the plate dry. This is illustrated in Figure 11-5.

Now we need to prepare the other electrode. To do this you will need another of the coated glass plates (the one with the conductive tin oxide coating—not the one with a titanium dioxide coating). You need to find which is the conductive surface. There are two ways of doing this—the tactile method is to simply rub the plate. It should feel rougher on the coated side. The other involves

a voltmeter or continuity tester. The conductive side is the one which yields a positive reading when tested for continuity.

We now need to deposit a layer of graphite. The easiest way to do this is take a soft pencil, and simply scribble on the surface until a nice even coating of graphite is obtained. This is shown in Figure 11-6. Just note that you need to do this with a plain pencil not a colored one!

Now if you have got this far, you are on the home run! The next thing we need to do is take some of the iodine/iodide mixture, and spread a few drops evenly on the plate that was stained with the dye (Figure 11-7). Once you have done this,

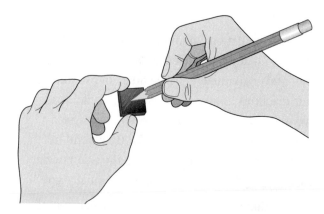

Figure 11-6   *Applying a graphite film to one electrode.* Image courtesy Greg P. Smestad.

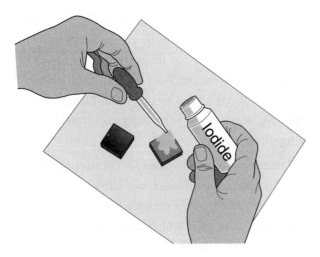

Figure 11-7 *Applying the iodine/iodide mixture.* Image courtesy Greg P. Smestad.

take the other electrode and place it on top of the dyed electrode. Stagger the junction between the two plates in order that you leave a little of each exposed at either end—you can then use a couple of crocodile clips to connect the cell to a multimeter.

Now clip the sheets of glass together carefully to ensure they stay together (Figure 11-8).

Now connect a multimeter—we can start to think about doing some really cool stuff now! You might like to try a few different experiments—like seeing what way to shine the light through the cell for the most effective operation. You might like to repeat some of the experiments in the section on photovoltaic solar cells, and see what results you obtain with a photochemical solar cell.

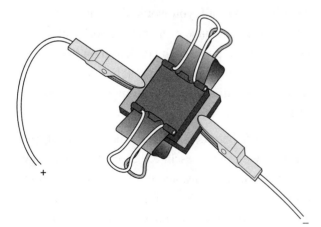

Figure 11-8 *Clipping the cell together with bulldog clips.* Image courtesy Greg P. Smestad.

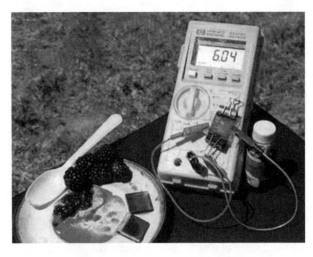

Figure 11-9 *The cell yields 6.0 mA.* Image courtesy Greg P. Smestad.

Another educational idea is to use a multimeter to measure the amount of power from both a photovoltaic solar cell, and the photochemical solar cell you have made, and compare the results—now work out their relative efficiencies taking into account the area of the cells.

Now we can take some measurements! Figure 11-9 shows a photochemical cell yielding 6.0 mA! Apparently, the juice in this picture is from Californian blackberries!

Figure 11-10 shows a photochemical cell being used to drive a small motor and fan.

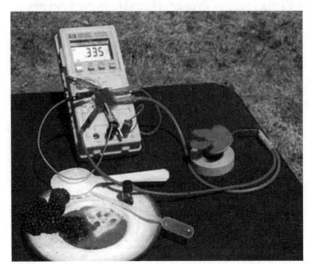

Figure 11-10 *The cell driving a small motor.* Image courtesy Greg P. Smestad.

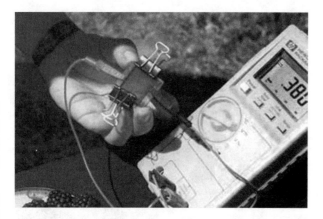

Figure 11-11 *A closeup of the photochemical cell in action.* Image courtesy Greg P. Smestad.

Figure 11-11 shows a close-up of the cell working in action!

Online resources

The materials required for this project are available from the Institute of Chemical Education from the following link or the address in the Supplier's Index: ice.chem.wisc.edu/catalogitems/ScienceKits.htm#SolarCell

Where does it all go from here?

This technology has a lot of promise for the future. There is a growing trend for manufacturers to integrate renewable energy systems into building elements—this allows us to feed two birds with one crumb, rather than shelling out for roof tiles and solar cells, why not buy a solar roof tile! The exciting thing about photochemical solar cells is that unlike photovoltaic cells, they don't necessarily have to be opaque. This opens up exciting possibilities—shaded windows and skylights which simultaneously produce electricity. How cool would that be!

When you consider all of the glazing that adorns the skyscrapers in our cities, you begin to realize that this technology has interesting applications for energy generation. It also allows us to make good use of daylight with our south-facing building areas, while generating energy at the same time.

There are also implications for consumer electronics, the watch giant Swatch has already built a prototype watch with a photochemical cover glass. This allows the glass which covers the watch to generate electricity all the time the watch is exposed to light. When you think that people wear watches on their wrists where they are permanently exposed to daylight, this becomes quite a sound idea! Of course, you also need some means of storing the electricity to enable the watch to run at night! It would be no good to wake up, put on your watch, only to find the time is set to the evening before.

Are there any limitations to this technology?

One of the problems with this particular type of cell is that the cell contains liquid which is essential for its function. Unfortunately, liquid is hard to seal and keep in—preventing the liquid from leaking is a real technical issue that needs to be solved. After all, you wouldn't want leaky windows! If you have ever seen a poorly fitted double glazing panel with condensation inside, you realize how hard it is to seal building fixtures and fittings against the ingress or egress of fluid.

However, there is hope on the horizon, Grätzel together with the Hoescht Research & Technology in Frankfurt, Germany, and the Max Planck Institute for Polymer Research in Mainz, Germany, have announced that they have developed a version of the cell with a solid electrolyte; however, efficiencies are low.

Photobiological solar cells?

Truth can sometimes be stranger than fiction. Realizing that conventional solar cells require expensive industrial processes, researchers at Arizona State University have initiated a project codenamed Project Ingenhousz which is looking at photosynthesis and how organisms can be used to harness solar energy to produce fuels that will wean us away from our carbon-based fossil fuels. Could your car one day run from hydrogen that has been produced by algae from solar energy?

Chapter 12

Solar Engines

In this book so far, we have seen how it is possible to utilize the energy that comes from the sun in order to do some really useful things. While generating heat and electricity is useful to help us reduce our energy consumption, it would also be useful if we could use the sun's energy to create mechanical movement. Mechanical movement is *very* useful and can be directly utilized to drive machinery.

When we look at the type of energy coming from the sun, we can see that it is heat and light—the energy is transmitted by radiation through the vacuum of space.

How therefore, do we exploit this radiated energy and convert it into mechanical motion?

You must be familiar with the steam engines that once graced railways throughout the world. The steam engine is a form of indirect combustion engine, the coal is burned to heat water, which undergoes a phase change—the water turns from liquid to gas. In doing so, it increases in volume—what once took up a small space, now takes up a large space—this change can be exploited to provide mechanical movement by driving a piston. Furthermore, the change in volume when hot

steam condenses back to water can also be used to provide movement.

If you want proof of this, take a soda can with a *little* bit of water in the bottom. Heat it on the stove until you see a little wisp of water vapor come from the can—this is evidence of the water having boiled. Now, using tongs, flip the can over, and immerse the "top" of the can in a bowl of ice cold water. The can is instantly crushed!

So now we have evidence that a change in temperature can produce movement, we can look at how to harness this raw power.

The engines described in this chapter, are all examples of thermodynamic heat engines. The chapter will showcase a few simple solar engines that you can construct yourself with relatively simple materials.

All of the engines in this chapter produce a fairly modest amount of mechanical power, but all serve to demonstrate that solar energy does have application in directly driving mechanical devices.

I gratefully acknowledge the advice and guidance of Hubert Stierhof in the preparation of a number of projects in this chapter.

Project 30: Build a Solar Bird Engine

You will need

- Happy Drinking Bird (won't be so happy when we are finished!)
- Silver spray paint
- Black spray paint

Tools

- Scalpel
- Kettle

We are going to have to "kill a bird" to execute this project—luckily the bird in question is

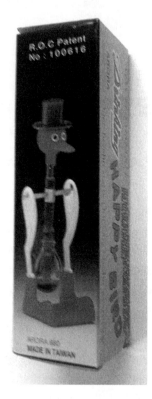

Figure 12-1   *The Drinking Bird in its box.*

Figure 12-2   *The Drinking Bird in action.*

Warning

The liquid which fills the Drinking Bird is known as dichloromethane. This chemical is pretty nasty stuff—safe while trapped in the glass of the drinking bird's body, but nasty if it escapes—so exercise caution, and try not to break or damage the bird in any way.

a cheap toy from the Far East, so our consciences are clear!

The "Happy Bird," "Happy Drinking Bird," "Drinking Bird," "Tippy Bird," "Sippy Bird," "Dippy Bird" or any one of a number of ridiculous names is a cheap novelty toy (Figure 12-1). It is also a serious scientific curiosity!

First of all a little note, don't jump straight into massacring the bird—have a little play with it, because it is a lot easier to see what is going on before you hack it (Figure 12-2).

How the Drinking Bird works

The Drinking Bird contains a pair of glass bulbs, joined by a glass tube, attached to which is a pivot. The pivot rests on top of the "legs" of the bird, and in the normal position, one bulb (the tail) will be full of liquid, while the other bulb (the head) will be empty.

The liquid inside the bird (dichloromethane) condenses readily with only a small temperature difference. In the normal mode of operation, you start the bird by "dunking" its head in a small glass of water, and then allowing it to stand. What happens when you do this, is that water begins to evaporate from the bird's head. As it does so, it takes heat with it, cooling the bird's head, relative to the ambient temperature, which the "bottom" bulb is at.

As the top is cooled, the dichloromethane gas in the head begins to condense creating a drop in pressure. The liquid from the bottom bulb now begins to flow into the head bulb.

As the liquid moves through the central tube, the weight distribution of the liquid in the bird changes. The bird now becomes "top heavy" and as a result of this, the bird rotates on its pivot. This movement is the useful "work" which the heat engine is doing.

As the bird tips over, the seal between the "neck" tube, and the surface of the dichloromethane is broken. A small bubble of vapor flows through the neck tubes. As it does so, it displaces liquid, which begins to flow back down to the tail bulb. As the liquid flows back from top to bottom, the vapor pressure equalizes and the bird "rights itself."

As the liquid flows back into the bottom bulb, the weight distribution again changes, and the bird pivots back to its vertical position. Again, more useful "work" is done here.

The cycle keeps repeating itself until the water in the pot runs out. The process is driven by the heat in the environment which causes the water to evaporate. The dichloromethane is not "used up" in any way—it is the working fluid of the engine and stays trapped inside the drinking bird.

We are going to change the Drinking Bird, so that instead of being driven by the temperature difference caused by water evaporating, it will instead be driven by the temperature difference created by surfaces that absorb and reflect the sun's rays.

How we are going to "hack" it . . .

We don't want to have to keep replacing our water in order to keep the bird working.

Now is the fun part, if you are of a destructive disposition.

We are going to take all of the little accoutrements away from this beasty! Boil a kettle and get some hot water, you will find it softens the glue and makes it a lot easier to remove things. Remember, the glass is only thin and easy to break.

Pull the hat of the bird. The hat usually disguises a little glass protrusion on the top of the bulb (Figures 12-3 and 12-4), so do this carefully, if you break the envelope of the bird it stops working.

Also, the little tail feather is going to need to go. Next you are going to have to remove the felt and

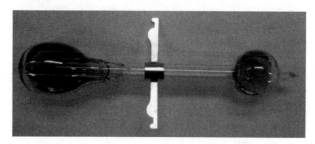

Figure 12-3 *The drinking bird stripped bare.*

nose from the head. With a sharp scalpel, you can cut through the plastic backing of the felt, and using hot water, you can scrape all the glue and gunk off. You should now be left with a nice clean piece of glassware.

Now, remember that the device works on the principle of temperature difference. You will remember from experiments earlier in the book, that black surfaces absorb solar radiation, while shiny or reflective light surfaces reflect solar energy. A black car feels hotter than the silver one

Figure 12-4 *The "hacked" drinking bird with it's fresh new paint scheme.*

next to it! So, get some spray paint—the type used for touching in dents on cars works well—and spray the bottom bulb black and the top one silver. Remember the evaporation of water cooling the "head" of the bird. Well, instead, reflective silver paint is going to keep the head cool. The black "base" of the bird is going to heat up as it absorbs solar radiation.

Now position the "solar engine" back on the legs of the bird, and put it somewhere where it will receive a lot of sunlight. You should now see the engine tipping away without the need for any water!

Project 31: Make a Radial Solar Can Engine

You will need

- Polystyrene ceiling tile or sheet polystyrene
- Three old cans
- Stiff wire (coat hanger wire is ideal)
- Three balloons
- Wooden strut

Tools

- Tin opener
- Scissors
- Wire snips

You might remember the biplane engines of old—they had their pistons arranged around a centrally driven shaft—the prop shaft—which would turn the propeller. This is known as a radial engine. In our cars, the pistons are generally arranged in a straight line or sometimes a "V," radial engines are different in this respect.

In this project, we will be building a radial engine, only rather than being fueled by aviation fuel, our engine is powered by the sun!

How does the solar radial engine work?

The cans that are exposed to the sun (i.e. not covered by the polystyrene shield) heat up as a result of the black covering absorbing the sun's rays. This increase in temperature results in the air inside the can expanding slightly. This increase in volume exerts a force on the rubber diaphragm covering the can. The diaphragm is connected to a short rod, which pushes against the crank turning the can assembly. When the can has rotated far enough so that it is covered by the polystyrene shade, then the sun's rays can no longer reach the can. As a result, the air inside the can begins to cool down. As the air cools, it contracts, in doing so, it pulls the rubber diaphragm (Figures 12-5 to 12-12).

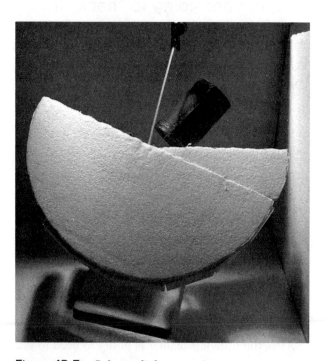

Figure 12-5 *Solar radial can engine.*

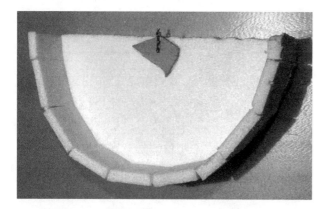

Figure 12-6   *The polystyrene shade.*

Figure 12-7   *The view from behind the can engine.*

Figure 12-8   *The can assembly (note the crank detail!).*

Figure 12-9   *The can assembly on its stand (from above).*

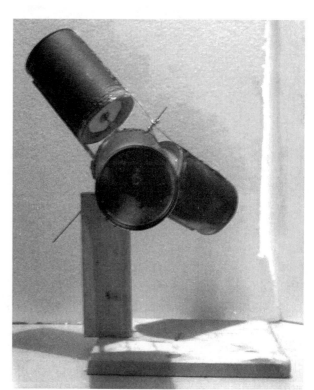

Figure 12-10   *The can assembly on its stand (from the side).*

Figure 12-11 *Note carefully how the crank is configured.*

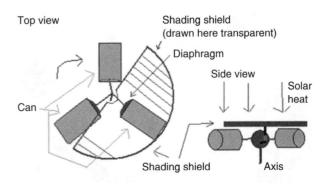

Figure 12-12 *Diagram of the radial can engine.*

Chapter 13

Solar Electrical Projects

In this chapter, we cover a number of small electronic projects you can build that are powered by solar energy. The chapter aims to show you how many common household devices that we take for granted could potentially run successfully on solar energy.

Project 32: Build Your Own Solar Battery Charger

You will need

- AA battery holder
- 9 V battery clip (you might need this to connect to your battery holder)
- 8 × solar cells 0.5V, 20 to 50 mA in full sun
- 1N5818 Schottky diode

Rechargeable batteries make good economic and environmental sense. In the same way that you wouldn't throw away the glass every time you had a drink, so it doesn't make sense to dispose of batteries when others are available that perform the same task many times over.

It gets even better than that when you realize that you don't have to use any mains electricity at all to recharge your batteries—you can use the power of the sun!

Solar battery chargers like this model are commercially available, which can be obtained from the Centre for Alternative Technology, U.K. (see Supplier's Index) (Figure 13-1); however, if you are electronically minded you can easily put one together.

The circuit here will recharge a pair of AA batteries quite happily when left in the sun.

This circuit is a very simple design which doesn't provide any regulation, so you will need to make sure that you disconnect your batteries when they are recharged.

The Schottky diode prevents the batteries' charge from flowing back through the solar cells when no charge is present. Schottky diodes have the advantage of not sapping too much of the power from our solar cells—maximizing the amount that is delivered to the batteries.

The schematic for the battery charger is shown in Figure 13-2.

Hint

If you live in a climate where sun is a rare treat and it is often overcast, you might like to experiment with a couple of additional cells in series to increase the power produced.

Construction is fairly simple. There are a wide array of cases available that are suitable for housing such a project. If you can get a housing with an integral battery holder you will find it will make neat work of housing the project.

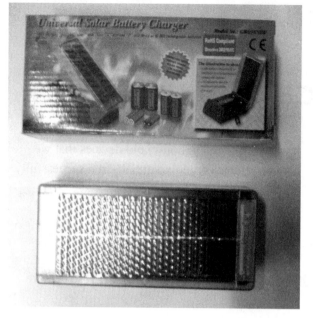

House the cells in such a manner that they are shaded and protected from the sun. If the cells get too hot their electrolyte leaks—damaging the cell and making a mess.

Figure 13-1 *Commercially available solar battery charger.*

If you want to build a "deluxe" model, you might want to consider incorporating a small milliammeter to monitor the charging status.

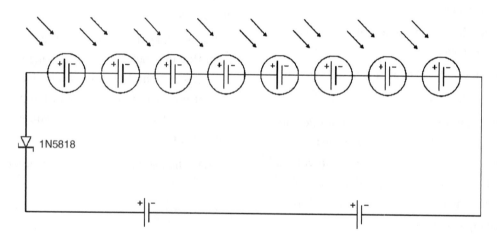

Figure 13-2 *Solar battery charger schematic.*

Project 33: Build Your Own Solar Phone Charger

You will need

- Car charger suitable for your cell phone (for cannibalization)
- 7812 voltage regulator
- 15 V solar array

Or for the USB version:

- 100 μF capacitor
- 100 pF capacitor
- 1 mH inductor
- 1N5819 diode

Figure 13-3 *A solar-powered phone—albeit not all that portable.*

- 274 k resistor (1% tolerance)
- 100 k resistor (1% tolerance)
- 100 µF capacitor
- USB socket
- MAX 630 CPA integrated circuit

It's the same old story—just when you want to talk on your cellphone, the battery goes flat and the conversation is irretrievably lost! Invariably, you haven't got your phone charger on you, and even if you did have it wouldn't be an awful lot of help as the chances are there is no power for miles around . . .

At the Centre for Alternative Technology, U.K., there is a solar-powered phone (see Figure 13-3); while this is powered by clean green energy, it can't claim to be very portable!

In this project, we are going to build a circuit that will provide a supply capable of powering either a cellphone or PDA charger. A PDA is about the limit of what you can charge using small cells, a laptop charger is probably a bit ambitious.

One of the problems with trying to build this circuit is that finding a suitable connector for many mobile phones is a real problem. While Nokia makes life easy by providing a simple jack that can be readily obtained from many component suppliers, many other manufacturers rely on proprietary connectors which are nonstandard and awkward to source.

For this reason, we have based this project on hacking a cellphone car charger.

There are two schematics here for projects that tackle the project from slightly different angles. The first method involves creating a solar array that will provide above 12 V—regulating this supply to 12 V, and charging the device via a hacked "car charger" (Figure 13-4). The other device is suitable for where a USB type charger is available—this is ideal for USB mp3 players, PDAs and mobile phones, most of which now come with a "data" lead. We have an array of solar cells, which charges a couple of batteries when there is spare power; a voltage regulator then turns this into a clean 5 V, which can be used to drive the device (Figure 13-5). The advantage of this circuit is that even if there is not a lot of sun— or it is night-time, you can pop a couple of freshly charged batteries in (maybe from your solar charger?) and things will start working.

Car chargers are designed to allow you to plug your phone into your vehicle's cigarette lighter or accessory socket. They are cheap and readily available; however, they rely on having a car present to allow you to charge your phone!

There are a couple of ways of making this project. You can either build the project as a box into which you plug your car cellphone charger, or, if you are a little more adventurous, you can take apart the cellphone charger and integrate it properly into the box. The plus side of keeping the two pieces separate is that you can use the car cellphone charger as a stand-alone item, or, you can power it from the "solar box." The plus side of integrating it all together is that it makes for a neat, stand-alone project and the two parts cannot become separated.

Note

A note on cigarette lighter sockets—the usual wiring scheme is that the casing of one of these sockets is connected to the negative terminal of the battery

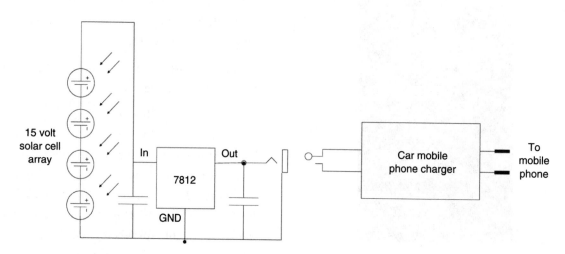

Figure 13-4 *Solar-powered phone charger schematic: car charger.*

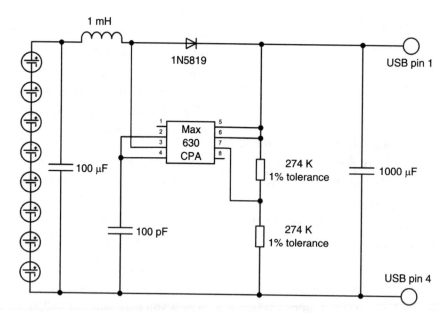

Figure 13-5 *Solar-powered phone charger schematic: USB type.*

Project 34: Build Your Own Solar-Powered Radio

You will need

- Ferrite rod antenna
- 60–160 pF variable capacitor
- BC183 transistor
- 10 nF capacitor
- 0.1 mF capacitor × 2
- 470 µF capacitor
- 220 R resistor
- 1 k resistor
- 100 k resistor × 2
- 10 k potentiometer
- Speaker
- PV cell
- PCB stripboard

Tools

- Soldering iron
- Solder

A solar radio is another great idea! Never mind "Desert Island Discs," with a solar radio, you can ensure that if you are ever marooned, you are able to listen to your favorite radio stations!

In this circuit we are going to build a simple AM radio that is powered by the sun.

There are some commercially available radios powered by solar energy; however, it is relatively easy to build your own. We are basing this circuit around the MK484 integrated circuit, which takes all the hassle out of building a simple radio. The integrated circuit looks like a transistor with three pins, and reduces the amount of external components needed considerably.

The schematic for the circuit is shown in Figure 13-6.

The radio has two controls. The variable capacitor changes the frequency that you are tuned to, and the potentiometer acts as a volume control for the simple transistor amplifier.

There are a number of commercially available solar radios; one idea for mounting, which you

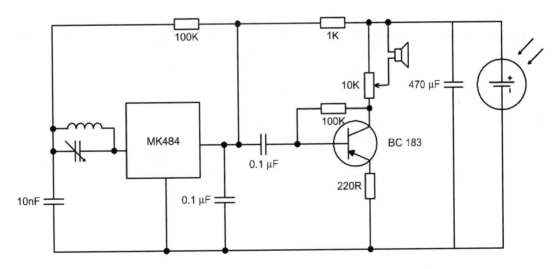

Figure 13-6 *Solar AM radio schematic.*

Figure 13-7 *Headphone mounted commercial solar radio.*

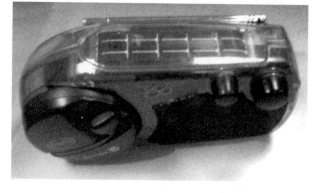

Figure 13-8 *"Freeplay" wind-up and solar radio.*

could easily accomplish with the solar radio circuit above, is to mount the circuit in a set of headphones, like the solar radio shown in Figure 13-7.

The radio in Figure 13-8 is the "Freeplay" wind-up radio invented by Trevor Bayliss. It uses two renewable energy sources—solar power, and for less sunny days "human wind-up power" in order to make sure that even when the sun doesn't shine, you aren't without your tunes!

Project 35: Build Your Own Solar-Powered Torch

You will need

- 4× 1.5 V solar cells
- 1× AA 600 mAh NiCad battery
- 1N5817 Zener diode
- 220 k ¼ W carbon film resistor
- 100 k ¼ W carbon film resistor
- 91 k ¼ W carbon film resistor
- 10 k ¼ W carbon film resistor
- 560 R ¼ W carbon film resistor
- 2× 3.3 R ¼ W carbon film resistor
- C9013 NPN transistor
- C9014 NPN transistor
- C9015 PNP transistor
- 300 pF ceramic capacitor
- 100 nF ceramic capacitor

- 1 nF ceramic capacitor
- 82 μH inductor
- CdS photocell 47 k @ 10 lux
- 2 × LEDs

In lists of made-up useless things, solar-powered torches seem to come out somewhere at the top. After all, what use is there for something that produces light that is powered by light? Until you realize that we can use *batteries* to store the energy—this is a crucial leap in understanding! Now doesn't the solar torch seem so much more interesting?

A solar torch is a useful thing to build and then leave on a sunny window sill. In the event of a power cut, you know that you can go to your trusty solar torch to provide a (somewhat modest) amount of illumination!

Figures 13-9 and 13-10 show a solar-powered torch and the solar torch in its packaging. One of the things you need to think about if you are going to house your project in a round torch case, is that you will need to ensure that either:

- The torch is weighted so that it rolls in such a way that the solar cell points upwards.

  or

- There is a flat machined into the case, which ensures that the solar cell points upwards when the flat in the case rests on a level surface.

Tragedy would strike if your solar torch were to roll over so that the flat faced away from the ground—blocking sunlight to the solar cell!

The circuit is shown in Figure 13-11. It is a variation of the outdoor solar light circuit (which you will see later in this chapter), where a pair of resistors and a switch are used to mimic the action

Figure 13-9   *Solar-powered torch.*

Figure 13-10   *Solar-powered torch in its packaging.*

of the photocell. It allows manual control of the LEDs and economizes by only using a single battery.

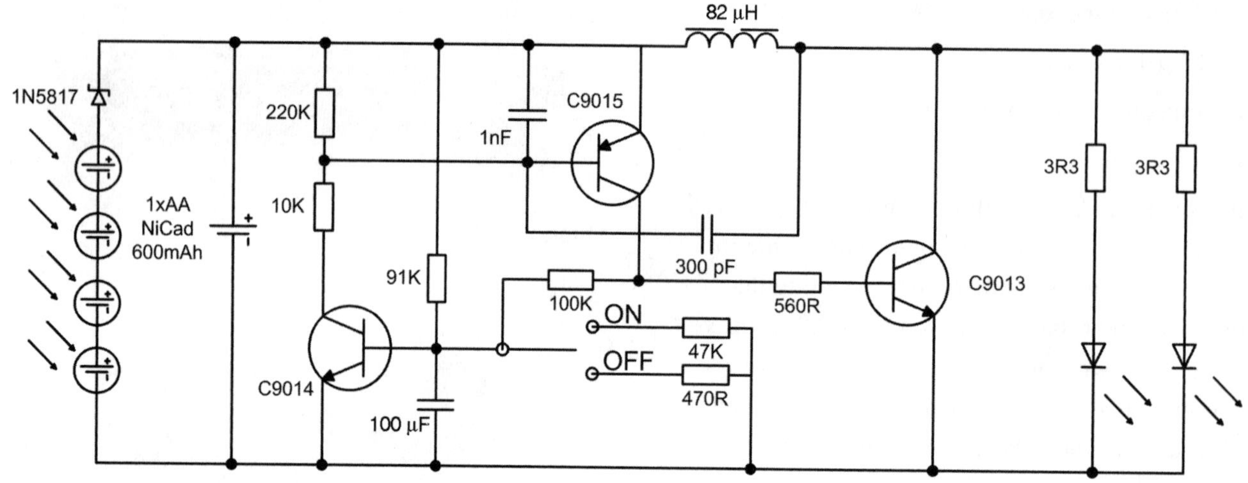

Figure 13-11 *Solar-powered torch schematic.*

Project 36: Build Your Own Solar-Powered Warning Light

You will need

- Capacitor 0.1 F, 5.5 V
- Capacitor 100 μF
- Capacitor 6.8 μF
- 2 × resistors 100 k
- 2 × resistors 100 ohms
- PNP transistor
- NPN transistor
- 2 × diodes 1N4148
- Super-high brightness red LED
- 100 μH inductor
- 4 × small solar cells

Tools

- Soldering iron

There are many applications where it is useful to have some sort of warning light, strobe, or beacon.

Often, the place where you want to position the warning light or strobe is totally remote from any source of power. Although we can often run things from batteries, sometimes we want to put a light where changing a battery would be undesirable. Solar energy, as well as producing clean renewable energy, also allows us to power things in remote places that would not easily be accessible using conventional cables, or where changing a battery could present a problem.

In Figure 13-12 we see a commercially available solar waterproof warning light, there are many applications for this—you might want to strap it to your back while cycling, for instance.

The beacon has a couple of modes. When the beacon is set to off, the solar cell will charge the battery; however, the light will not flash under any circumstances. In "solar" mode, the beacon will charge during the day, and when the circuit senses a low lighting condition, the beacon will begin to flash using the power stored in the rechargeable battery. Set to "on" the beacon will flash regardless

Figure 13-12 *Solar waterproof warning light.*

of whether it is light or dark—however, bear in mind that this will drain the battery.

If you are going to use this beacon outside all of the time, you might want to think about how you can protect the circuit (Figure 13-13) against the ingress of water and solid matter. Most suppliers of cases sell a range of decent waterproof cases that are eminently suitable for outdoor use, or you may find that you can improvise with a Tupperware or similar container to produce a satisfactory housing.

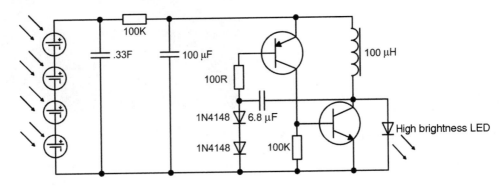

Figure 13-13 *Solar waterproof warning light schematic.*

Project 37: Build Your Own Solar-Powered Garden Light

You will need

- 4 × 1.5 V solar cells
- 1 × AA 600 mAh NiCad battery
- 1N5817 Zener diode
- 220 k ¼ W carbon film resistor
- 100 k ¼ W carbon film resistor
- 91 k ¼ W carbon film resistor
- 10 k ¼ W carbon film resistor
- 560 R ¼ W carbon film resistor

- 2 × 3.3 R ¼ W carbon film resistor
- C9013 NPN transistor
- C9014 NPN transistor
- C9015 PNP transistor
- 300 pF ceramic capacitor
- 100 nF ceramic capacitor
- 1 nF ceramic capacitor
- 82 µH inductor
- CdS photocell 47 k @ 10 lux
- 2 × LEDs

Tools

- Soldering iron

Solar-powered path lights (Figure 13-14) are becoming ubiquitous in just about every garden center nowadays! There are lots of advantages to using solar power rather than a hard-wired system. First of all, as a hard-wired system is exposed to the elements, you need to ensure that you use low-voltage fixtures and fittings, which require a transformer to step down the voltage, or failing that, really expensive mains fixtures and fittings. Then the next thing to consider is that even the safest low-voltage system is still vulnerable to the gardener's spade—a badly placed spade can mean disconnection of your garden lighting system.

Solar-powered garden lights have none of these disadvantages. They charge their batteries during the day, and then at night as the light fades, they switch on, providing illumination.

The change in illumination is detected by a CdS photocell.

We will be using LEDs for this project (Figure 13-15) as they provide good efficiency—a decent amount of illumination for the relatively small amount of energy we are able to provide.

Figure 13-14   *Solar garden light.*

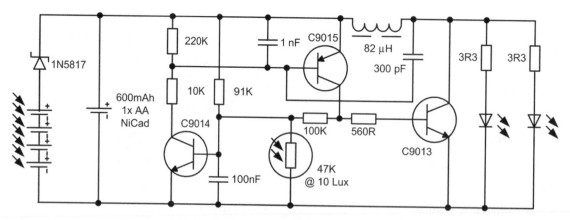

Figure 13-15   *Solar garden light schematic.*

Chapter 14

Tracking the Sun

One school of thought advocates positioning your solar panels on a fixed surface such as a roof, positioned so as to harness as much sun as possible *on average* over the year. This approach certainly works, but as we saw in Chapter 3, the sun is not a fixed object in the sky—it moves, and so this approach is not necessarily the best.

One other solution is to actively track the sun using a device such as the trackers shown in Figures 14-1 and 14-2. What this entails is using motors, hydraulic actuators, or some other devices to move our solar panels to follow the sun. This approach does have some merits. With the sun always facing the panels as face-on as possible, the most possible energy is extracted, as the panels are operating at their greatest efficiency.

One of the main caveats of this design is that moving the panels does require some input of energy, and this must of course be subtracted from the total energy that the panels are producing.

Furthermore, in some scenarios this is inappropriate—if you are roof-mounting panels, it would not really be appropriate to "reposition your roof" every time the sun moves.

In this chapter we will build a circuit that can be used to track the sun's movements and move a solar panel accordingly. The circuit is only simple, and will power a small motor to drive a demonstration display; however, with correct driver circuitry, the circuit could easily be scaled up to move bigger arrays.

Figure 14-1   *Solar tracker at Llanrwst, near Snowdonia, Wales.* Image courtesy Dulas Ltd.

Figure 14-2 *Solar tracker at Llanrwst, near Snowdonia, Wales.* Image courtesy Dulas Ltd.

Tip

If you don't fancy building your own solar tracker from scratch, Science Connection sells one as a kit under the stock code number 2216KIT. Full details are in the Supplier's Index.

www.scienceconnection.com/Tech_advanced.htm

Project 38: Simple Solar Tracker

You will need

- 3 × LDR
- 33 R resistor
- 75 R resistor
- 100 R variable resistor

- 10 k variable resistor
- 20 k variable resistor
- 2N4401 transistor
- TIP120 Darlington pair
- 9 V relay
- 5 V motor

How the circuit works

We have three CdS photoresistors (Figure 14.3) The value of these resistors is about 5 k when not exposed to light. However, when we expose them to light their resistance decreases to around about a couple of hundred ohms.

The third CdS cell is mounted in a shrouded enclosure so that it is only illuminated when it faces directly toward the sun. When the sun illuminates this photocell, its resistance drops, and as a result, when the sun shines on it, our Darlington pair is kept off.

When the sun moves away from the line of sight of photoresistor 3, its resistance increases. This allows our Darlington pair to switch on, which in turn drives our relay, which in turn drives our motor to move the array.

The variable resistor in line with the relay and motor allows us to regulate the motor's speed. The motor should turn slowly enough to move the array, but not so fast that the array overshoots before photoresistor 3 has a chance to respond to the change.

Photoresistor 2 is mounted flush with the panel so that it can see the whole sky. Its function is to check that the sun is present, to prevent the array from searching for a sun that isn't there! If the sun is present it will sense this and drive the base of our NPN transistor low. However, if the sun

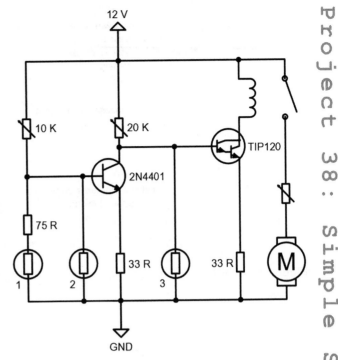

Figure 14-3 *Simple solar tracker schematic.*

disappears behind a cloud its resistance rises and our NPN transistor base is allowed to be high, this in turn drives the base of our Darlington pair low, which prevents the tracker from tracking.

Our first photoresistor is mounted on the back of our tracker. It senses the new light coming from the east and activates the turntable to allow our solar panels to face the sun to catch the new light.

The setup for our solar array and sensors is illustrated in Figure 14-4.

Hint

If you want to adjust the sensitivity of a CdS photoresistor without fiddling with the electronics, you can decrease its sensitivity to light by drawing on a section of the sensing element with a black permanent marker. This prevents some light reaching the sensor.

Tip

Poulek Solar, Ltd., whose website is www.solar-trackers.com sell commercially produced solar tracking circuits and all of the hardware to mount your panels on a sturdy outdoor tracker. See Supplier's Index.

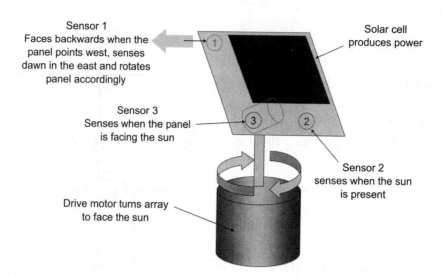

Sensor 1
Faces backwards when the panel points west, senses dawn in the east and rotates panel accordingly

Solar cell produces power

Sensor 3
Senses when the panel is facing the sun

Sensor 2
senses when the sun is present

Drive motor turns array to face the sun

Figure 14-4 *Solar array and sensors setup.*

Taking it further

You don't have to use this simple circuit just to move a solar panel, you can think of ways to move any of the solar projects presented in this book. You might want to move a solar cooker for example. The motor used in the circuit might only be small; however, you can use gearing to enable it to move larger loads. We only want slow motion from our motor anyway, so slow movement is ideal.

Online resources

If you want to explore more sophisticated solar-tracking devices, here are some links to hobbyists pages which will take you further in your design of solar trackers.

pages.prodigy.net/rich_demartile/

www.redrok.com/electron.htm#tracker

www.phoenixnavigation.com/ptbc/articles/ptbc55.htm

Solar trackers in the real world

Now we have built a model, let's take a look at a *real* solar tracker and gain some insight as to the capabilities of the technology.

Mr Howie, Scotland

Here, it was decided that solar trackers were the way forward, as the roof structure of the house on the property was not strong enough to support a solar array, and the rafters were irregularly spaced—making it difficult to install mounting hardware. As the property was surrounded by a lot of land, it was decided that it was cogent to install a stand-alone array. As Scotland is at quite a high latitude, it was decided that using a solar tracker would make the best use of the available solar resource.

The array (Figure 14-5) is 1.92 kW peak, and was 48% funded by an Energy Saving Trust grant.

Figure 14-5 *Solar tracker on the property of Mr Howie, Scotland.* Image courtesy Dulas Ltd.

Chapter 15

Solar Transport

Why solar transport?

The way that we live today necessitates traveling long distances—whereas historically all travel was on foot, we are now assisted by cars, boats, and trains to get us from place to place.

Our world has shrunk—low-cost air travel now means that we can be anywhere in the world affordably within the day, and the car means that we can travel almost anywhere local by road in a matter of minutes.

Our world today has been styled and shaped by our transport patterns. Many people live in the suburbs and commute for what would have been incredibly long distances in the past to get to their places of work, shops, and amenities.

Where in the past shops were a local affair, now we go to large sprawling out-of-town shopping centers and malls.

All of this increased transport affords us endless convenience, but what is the real cost?

The environmental cost of transportation

The city of Los Angeles in the U.S.A. is an example of a city that has learnt to pay the price for heavy use of transportation. The city's urban planning has dictated that people use their own private vehicles, as public transportation is poor.

Transportation uses the bulk of the world's petroleum. We use petrol or gas (depending on which side of the pond you come from), because it is an energy-dense, readily-available (at present) fuel, which provides power-on-demand when we want it.

However, imagine a world without cheap gasoline . . . How would we get about? As Chapter 1 mentioned, a world without petroleum may be here sooner than we think.

Furthermore, the burning of lots of gasoline and diesel results in all sorts of "nasties" going into the air that you and I breathe—this includes carbon dioxide, oxides of sulfur which are responsible for acid rain, oxides of nitrogen, particulates, and unburnt hydrocarbons. We are putting this lethal cocktail into our air day-by-day.

What are the alternatives?

Well, for a start we can try to change our transport patterns. Social fixes like this are really cheap and necessitate minimal investment. What this means in practice is drive less and fly less. It might sound tough, but in fact it really is easy to make a conscious effort to reduce our transport patterns.

Also, in addition to trying to reduce the amount that we travel, we can try to reduce the amount that other people have to travel. This could be done, for example, by sourcing locally produced products.

We can try and reduce our carbon emissions by using public transportation—it follows that it is more efficient to move a large number of people than a small number of people—the efficiency gains and economies of scale mean that we save fuel and avoid many emissions.

Figure 15-1 *Honda dream car. Image courtesy Honda.*

However, there must be alternatives to our present fossil fuel-based vehicles, and yes there are. You'll never guess what . . . these alternatives are solar derived as well! Read on and see what the exciting technology has in store.

Solar vehicles

At the moment, solar cars aren't really practical for you and I to drive about in—simply too much surface area is required to mount the solar cells. Also, some method of storing the solar energy needs to be employed for when your car goes into a tunnel or the sun hides behind a cloud. Even so, solar vehicles such as the Honda Dream shown in Figure 15-1 are an interesting demonstration that it is possible to produce a vehicle that runs on solar energy.

There are a number of competitions that aim to spur on the development of solar vehicles, two notable competitions are the World Solar Challenge, and the North American Solar Challenge. If you are really keen on getting into solar vehicles, some of the top universities enter cars into the races.

OK . . . so entering a full-size solar car into a race is a little bit pricey (Figure 15-2), but what can you do instead to fulfill your solar ambitions? Follow the next project and find out how to build a simple solar vehicle!

Online resources

Check this website about solar vehicles for some cool information

www.formulasun.org/education/seles9.html/

Online resources

Take a peek at the World Solar Challenge website to see what is going on

www.wsc.org.au/

Also, the North American Solar Challenge is here

www.americansolarchallenge.org/

Figure 15-2 *Solar vehicles lined up for a solar car race.* Image courtesy NASA.

Project 39: Build Your Own Solar Car

In this project we will be building a small solar vehicle that demonstrates how solar power can be used to propel a small vehicle. In the next projects, you will learn how to "soup up" your racers and race them.

You will need

- SolarSpeeder 1.1 printed circuit board (PCB)
- High-efficiency coreless motor
- Motor mounting clip
- 3 × Rubber wheels on nylon hubs
- 43 mm long 1.40 mm diameter (1.7 in. long, 0.055 in. diameter) steel rod
- 2 × Black plastic wheel retainers
- 0.33 F 2.5 V gold capacitor
- 2n3904 transistor

- 2n3906 transistor
- 1381 voltage trigger
- 2.2 k resistor (color bands red/red/red/gold)
- SC2433 24 × 33 mm 2.7 V solar cell
- Pair solar cell wires
- 25 mm (1 in.) length 18 gauge wire

Tools

- Soldering iron
- Needle-nose pliers
- Side-cutters or strong scissors
- File and/or sandpaper
- Glue, rubber cement, or hot-glue (or superglue, if you're very careful)
- Safety glasses—*very* important when clipping and snipping!

First, assemble all of the components for the Solaroller as in Figure 15-3.

When our project is complete, it will look like the pretty little bug in Figure 15-4.

First of all, take a look at the schematic in Figure 15-5. This is a pretty standard design for a solar engine. What is happening here is that our

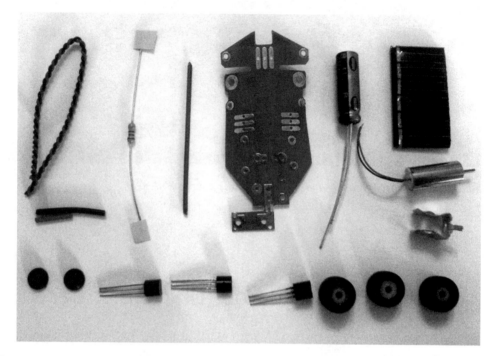

Figure 15-3   *Components of the Solaroller.*

Figure 15-4   *The assembled Solaroller.* Image courtesy Solarbotics.

little solar cell is providing electrical energy which is charging the high-capacity capacitor. When the voltage reaches a certain threshold level, the 1381 triggers the output circuit, which dumps the power in the capacitor through the motor, creating movement.

The first step of assembling your Solaroller is illustrated in Figure 15-6. You are going to need to take the axle and thread it through the two holes in the circuit board named "rod."

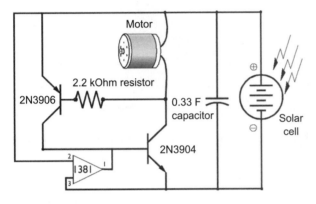

Figure 15-5 *The Solaroller schematic.* Image courtesy Solarbotics.

Next, take the high-capacity capacitor, bend the leads so that they are flush with the body of the capacitor. Then solder it into the PCB. Ensure that you solder this in the correct orientation.

Next, take the 2.2 k resistor and solder it as shown. The orientation of the resistor is unimportant.

The next stage in assembly is shown in Figure 15-7.

First take the 3904 transistor and solder it at the head of the board in the orientation shown in Figure 15-7.

Now the 1381 and 2906 transistors are soldered in either side of the board facing down. This is also illustrated in Figure 15-7.

Finally take the small fuse clip which will be acting as our motor mount and solder it into the bottom of the board. Note that the fuse clip has a small lip to prevent the motor sliding out. Ensure that you orient this correctly.

Now you have got this far you are definitely cooking with gas! . . . or should that be with solar? Now take the small high-efficiency motor and insert it into the fuse clip in the manner shown in Figure 15-8.

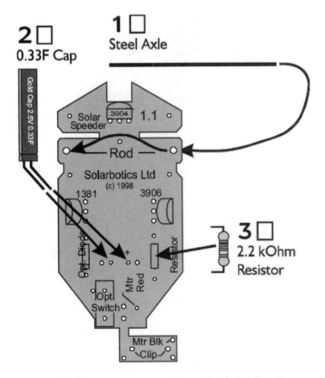

Figure 15-6 *Step 1—assembling the* Solaroller. Image courtesy Solarbotics.

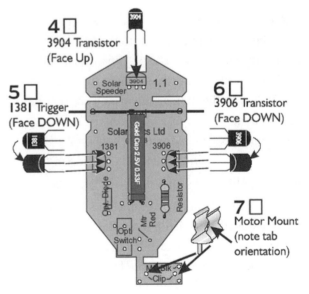

Figure 15-7 *Step 2—assembling the Solaroller.* Image courtesy Solarbotics.

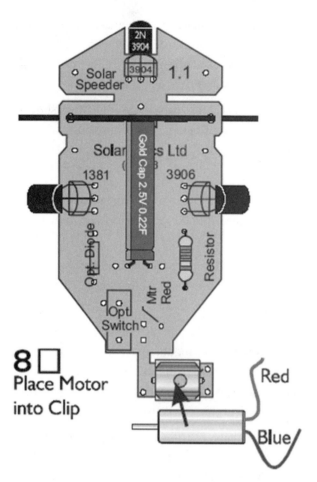

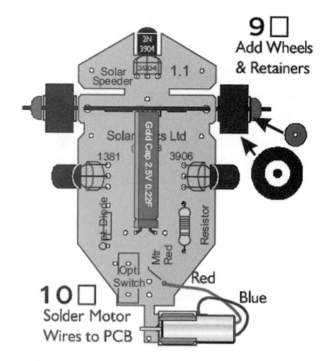

8 ☐
Place Motor into Clip

Figure 15-8 *Step 3—assembling the Solaroller.* Image courtesy Solarbotics.

9 ☐
Add Wheels & Retainers

10 ☐
Solder Motor Wires to PCB

Figure 15-9 *Step 4—assembling the Solaroller.* Image courtesy Solarbotics.

Adding the wheels at the front is a simple procedure of pushing them onto the axle and then adding the small black plastic clips which will retain the wheels and prevent them from sliding off. Now take the motor leads which are *very* delicate so treat them with a lot of respect! The red one should be soldered into the hole on the PCB, the blue one should be soldered onto one of the holes near the fuse clip (Figure 15-9).

Next take a small piece of thick copper wire and separate the insulation from the copper wire (making sure that you keep the insulation intact as we will be needing this later!). The wire should be bent at one end and soldered first into the hole adjacent to the motor clip, and then to the motor clip itself to provide mechanical support (Figure 15-10).

TOP VIEW

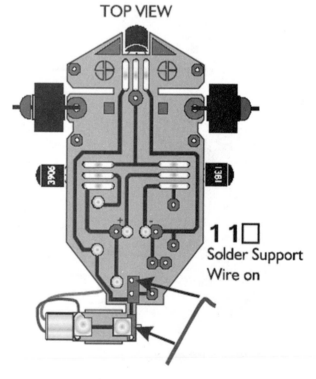

11 ☐
Solder Support Wire on

Figure 15-10 *Step 5—assembling the Solaroller.* Image courtesy Solarbotics.

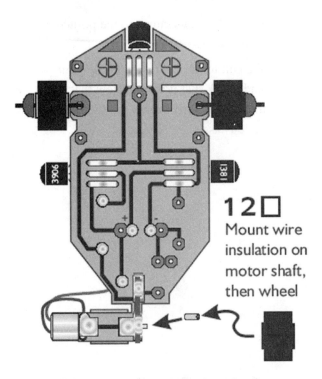

12 ☐
Mount wire insulation on motor shaft, then wheel

Figure 15-11 *Step 6—assembling the Solaroller.* Image courtesy Solarbotics.

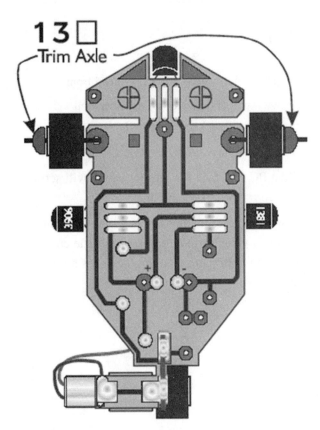

13 ☐
Trim Axle

Figure 15-12 *Step 7—assembling the Solaroller.* Image courtesy Solarbotics.

14 ☐ Prepare Solder Pads

Bare Pad

Properly Tinned Pad

Figure 15-13 *Step 8—assembling the Solaroller.* Image courtesy Solarbotics.

15 ☐ Glue Wires Down

Glue wires down only in this area

Figure 15-14 *Step 9—assembling the Solaroller.* Image courtesy Solarbotics.

Next, take a short length of that insulation that you saved and slide it onto the motor shaft. Now take the wheel and slide it over the insulation (Figure 15-11).

The next step is dead simple! Trim the axles at the front of your Solaroller (Figure 15-12).

Now tin the pads on the back of the solarcell (Figure 15-13).

Now solder the wires to the tinned pad, and add a little dab of glue in an area away from the soldered joints to act as a strain relief (Figure 15-14).

16 ☐ Connect Solarcell to PCB

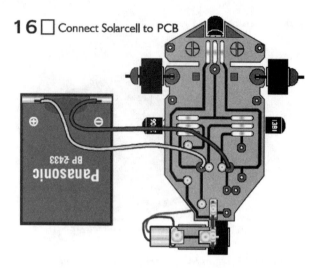

Figure 15-15 *Step 10—assembling the Solaroller.*
Image courtesy Solarbotics.

Now solder the connections to the printed circuit board (Figure 15-15).

Hold the cell to the light, supporting your solar vehicle and check that it works. Now that you have proved that the circuitry works, fix the solar cell to your vehicle chassis.

Project 40: Hold Your Own Solar Car Race

You will need

- A stopwatch

    or

- Lap timer 200 software and a PC

- A number of solar cars

OK, so the World Solar Challenge, and the North American Solar Challenge might be a little out of your reach; however, holding your own tabletop solar car race certainly isn't.

The free lap timer software presents a high-tech alternative to simply using a stop watch to time your cars. The software comes with schematics to

Tip

Lap timer software is a free download from:

www.gregorybraun.com/LapTimer.html

build a PC interface for sensors which will sense when your car crosses the line.

You might like to consider how you can make your team vehicles look different. A little customization with paint and graphics goes a long way!

Project 41: Souping Up Your Solar Vehicle

- Think about how you could make a solar concentrator with tin foil or Mylar reflective surfaces channeling more energy to your solar cell.

- Experiment with different tire types. You might find that some different wheels from another model car offer better grip.

- Try replacing the front wheels with some kind of skid. Think about reducing the Solaroller's friction; however, as you reduce friction, you might also

reduce control or the model's ability to travel in a straight line!

- You can tweak the value of the 2.2 k bias resistor. This will change the efficiency of the solar engine. Higher values will make your solar engine more efficient, but will increase the time taken to charge. Smaller values will speed the rate at which the motor is triggered, but at the expense of efficiency.

Project 42: Supercharge Your Solaroller

Additionally, you can add a diode which allows your Solaroller to charge more than would normally be possible. You have the option of using a bog standard glass diode or an LED.

The first thing you will need to do is cut through the PCB trace in Figure 15-16.

The next thing to do is take your diode, and solder it as shown in Figure 15-17. Make sure that you orient the stripe on the diode or flat on the LED correctly.

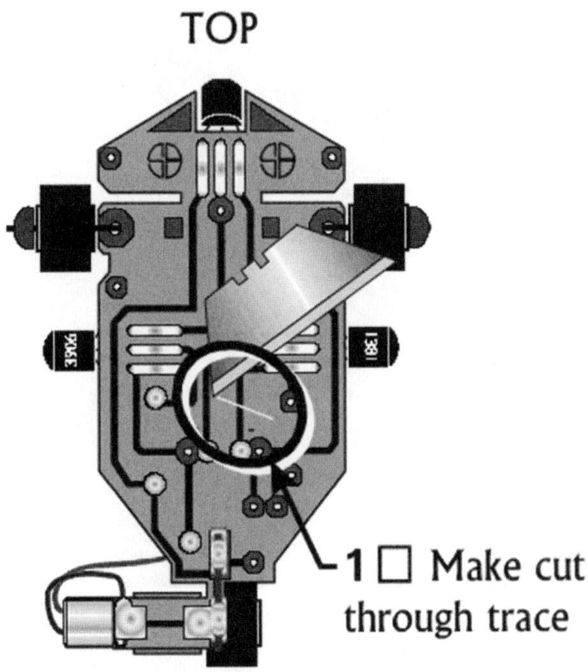

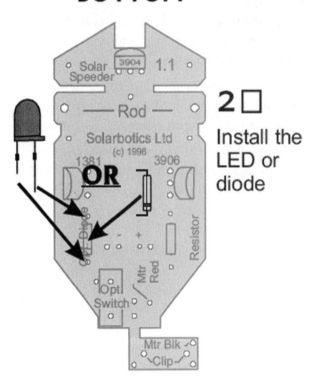

Figure 15-16 *Cutting through the PCB trace.* Image courtesy Solarbotics.

Figure 15-17 *Adding the diode.* Image courtesy Solarbotics.

Figure 15-18 *Honda fuel cell vehicle.* Image courtesy Honda.

Fuel cell vehicles

In Chapter 17, you will learn about hydrogen fuel cells. Hydrogen isn't a fuel as such—you can't "dig it out of the ground," and although it is the most common element in the universe, we can't access it in a readily usable way.

However, we are surrounded by water which is H_2O, this means that it contains two parts of hydrogen for every part of oxygen.

As you will see in Chapter 17, it is relatively easy to separate the hydrogen from the oxygen by passing an electric current through the water. There are other ways of producing hydrogen, but whatever the method, this hydrogen can then be used as an *energy carrier* to provide power for a fuel cell vehicle, such as the Honda FCX in Figure 15-18.

Solar aviation

Figures 15-19 to 15-21 show examples of solar-powered flight. Figure 15-22 shows a fuel cell.

NASA Dryden Flight Research Center Photo Collection
http://www.dfrc.nasa.gov/gallery/photo/index.html
NASA Photo: EC98-44803-29   Date: November 1998   Photo by: Tom Tschida

Centurion in Flight with Internal Wing Structure Visible

Figure 15-19   *Centurion solar plane.* Image courtesy NASA.

Figure 15-20   *Helios solar plane.* Image courtesy NASA.

Figure 15-21 *Helios 2 solar plane.* Image courtesy NASA.

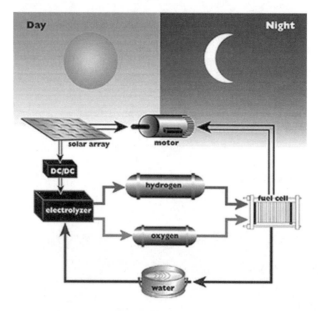

Figure 15-22 *Helios fuel cell diagram.* Image courtesy NASA.

Project 43: Build Your Own Solar Airship

You will need

- Solar airship tube (8 m of thin plastic tube)
- 2 × cable ties
- Tether line (50 m)

This project is one of the most ridiculously simple projects in this book, yet it is also one of the most visual and almost counterintuitive.

You might be under the misapprehension that to get things into the air you need sophisticated jet engines or rocket thrusters. Certainly if you have

Note

The kit for this project is well worth the spend, if only for the enormous scale of the airship—it is a massive 8 meters long. However, if you want to experiment, you could get some good results with cheap black plastic bin liners—why do I say cheap? Well cheap bin liners tend to be made out of thinner plastic and so are lighter for the amount of air they enclose. As they are already sealed at one end, you can get away with one cable tie for one end, and some fine fishing line for the tether.

Figure 15-23 *Solar airship.*

read my other book *50 Model Rocket Projects for the Evil Genius* (intentional plug) you will know all about rocket motors and what they can do—but hang about! There are also much simpler ways of getting things to fly, and believe it or not, they involve solar energy!

The procedure for flying a solar airship is simplicity itself. Take the long plastic tube. Put a cable tie around one end. Then, holding the other open, run until the tube is full of air. Try and fill it as much as you can and then bunch up the end and tie it tight with another cable tie. Attach the tether line to one of the cable ties.

Now, place your solar airship in the sun and watch what happens.

Gracefully, slowly, you will see your solar airship begin to twitch a little, and then rise into the air. Hang on to that tether—else it might escape you (Figure 15-23)!

As the airship ascends into the sky, you might like to take a few moments to think about what is happening here.

Look at hot air balloons—they are lifted by burning gas, a hydrocarbon, but what actually happens is that the gas is heating the volume of air inside the balloon. As the air is heated, it becomes less

Online resources

Solar airship links

Here are a number of places on the web where you can buy your own solar airship tube.

www.eurocosm.com/Application/Products/
Toys-that-fly/solar-airship-GB.asp

www.amazon.co.uk/exec/obidos/ASIN/
B000279OKI/202-9916112-9335805

www.find-me-a-gift.co.uk/gifts-for-men/
unusual-gadgets/solar-airship.html

www.comparestoreprices.co.uk/novelty-gifts/
unbranded-8m-solar-air-ship.asp

This experiment has got to be seen to be believed! If you want to check out someone else doing it first, you might like to take a look at this video on the net.

www.watchondemand.co.uk/solar-airship.htm

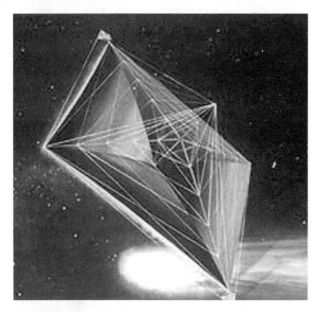

Figure 15-24   *Could solar sails be used for space travel?* Image courtesy NASA.

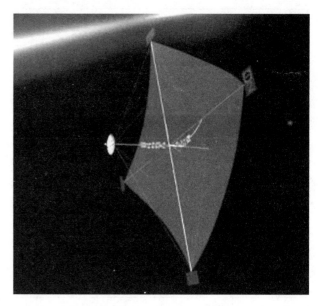

Figure 15-25   *A solar sail unfurled in flight.* Image courtesy NASA.

dense and therefore wants to float above denser air—this provides a lifting force. Simple really!

What does the future hold?

Space travel has long been a dream of futurists and science fiction lovers alike. However, traversing the voids of space presents a number of problems—one of which is where does the energy come from to propel a vehicle all that distance through space?

One possible answer could come in the form of a solar sail (Figure 15-24).

The sail shown in Figure 15-25 would be estimated to be half a kilometer across! Although the push provided by solar radiation is small, being harnessed over such a large area it is strong enough to provide propulsion.

Chapter 16

Solar Robotics?

I would like to thank Dave Hrynkiw at Solarbotics Ltd for his help in preparing this chapter.

Over the past century, we have seen a relentless march toward a world of automation and ease. Ever since the industrial revolution, increases in efficiency have been made by using automatic devices and robots to take over the menial tasks of man.

As a result, we have given ourselves over to the machine, and our modern world is dependent upon the functioning of these devices for our continued development and growth.

This presents us with a little bit of a dilemma.

The machines and automation of the industrial revolution were fueled by coal. More recently, oil, natural gas, and other fossil fuels have powered the wheels of industry and automation. While automation requires less human input, it does require energy.

Thus we have enslaved ourselves to previously plentiful fossil fuel energy, and created a world which would be much changed without it.

In the field of energy and robotics, we could then go on to say that if we want to push the boundaries of what we know, into the deep uncharted realms of space, exploring other planets (Figure 16-1) (or closer to home, remote inaccessible areas like the sea), we need to provide energy for these distant ventures—this is hard with conventional means.

Figure 16-1 *Spirit Mars rover.* Image courtesy NASA.

The Spirit Rover which was sent to Mars, was equipped with a 140 W array.

We see the increasing penetration of robots into our households—Roomba and Scooba are household names and represent readily available domestic robots. However, at present, service robots need clumsy recharging stations. What if your domestic robots could roam freely around your house powered by nothing more than the sun shining through your window, and the ambient light in your rooms?

BEAM robotics

BEAM (Biology, Electronics, Aesthetics, and Mechanics) robotics differ from traditional robotics in one important respect—whereas conventional robotics tends to employ a central processor and some software to dictate the behavior of the robot, BEAM robotics has a different approach. The robotic behavior is governed by simple circuits which interact with each other in defined patterns.

Note

The great news is that all of the above is available in a kit from Solarbotics.

Check out www.solarbotics.com/ for more details.

There is a coupon at the back of this book which will help you secure a discount on Solarbotics products.

The Photopopper Photovore

The Photopopper Photovore is a nippy little robot powered only by solar energy. While very simple in nature, it demonstrates that solar-powered devices can perform simple autonomous functions—this sets a precedent for much larger, more complex devices.

The circuit that the Photopopper Photovore uses is called a "Miller Engine" circuit (Figure 16-2), the solar cell charges a power capacitor, which stores the power ready to be used in bursts by the motor. The circuit uses a 1381 voltage trigger to trip the motor circuit, once the solar cell has charged the capacitor sufficiently. Once the capacitor has discharged beyond a certain point, the 1381 cuts off, and stops the motor—allowing the capacitor to charge. While you can use this circuit as a basis for your own designs, a full kit of parts for this robot is available from Solarbotics (see the Supplier's Index).

Photopopper behaviors

Although simple, we see a number of emergent behaviors as a result of the simple interconnections of the above circuit.

The first behavior that we see, is a light-seeking behavior. This means that our Photopopper will try to seek light and avoid shadow wherever possible. This behavior was exhibited in Grey Walter's turtle, an early robot which proved that a limited number of connections could give rise to more complex emerging behaviors.

We can see the way the robot makes its way toward a light source in Figure 16-3.

When we try and relate our mechanical robot beastie to the world of the animal kingdom, we can see that light is the "food" for our robot—a natural animal instinct is to go where there is food in order to survive—it is clear that this robot exhibits this behavior trait.

Another behavior that our robot exhibits is "obstacle avoidance" behavior, where the robot tries as far as possible to avoid obstacles by using its "whisker" touch sensors (Figure 16-4). Again, we can relate this to the animal kingdom, where whiskered animals such as hamsters and cats sense obstructions with their sensitive whiskers and take evasive action to avoid those obstacles.

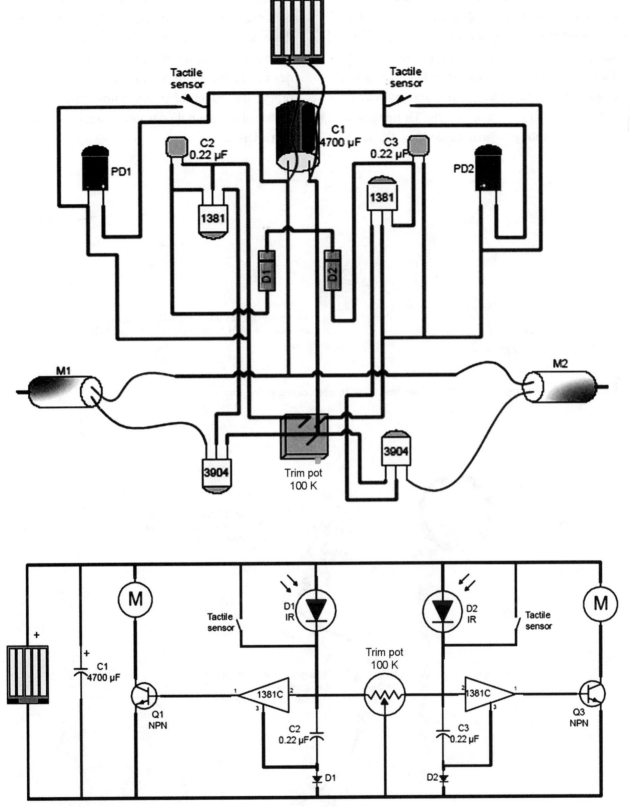

Figure 16-2 *Photopopper Photovore schematic.* Image courtesy Solarbotics.

You might say this is a very nice novelty, but where is the practical application? Sensing light and touch is an analog for sensing any number of variables. Imagine a robotic Hoover that would vacuum the carpet of your room, but not when it touched a wall or a piece of furniture, powered only by the sun and ambient light. Or imagine a robotic lawnmower which would mow the lawn, but not where it sensed that there were borders full of flowers or garden ornaments. As you can see, these behaviors, which on the face of it are very simple, can be combined to make an automated device that has more complex behaviors, and can relieve the tedium of simple human tasks, while at the same time not consuming precious fossil-fuel energy.

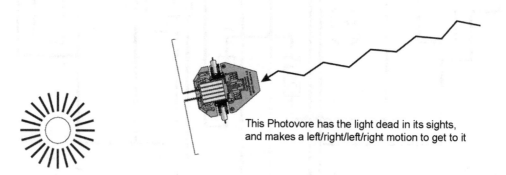

This Photovore has the light dead in its sights, and makes a left/right/left/right motion to get to it

Figure 16-3 *Photopopper light-seeking behavior.* Image courtesy Solarbotics.

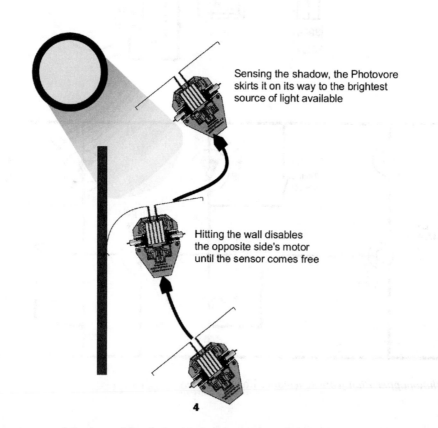

Sensing the shadow, the Photovore skirts it on its way to the brightest source of light available

Hitting the wall disables the opposite side's motor until the sensor comes free

Figure 16-4 *Photopopper obstacle-avoiding behavior.* Image courtesy Solarbotics.

Solar Robotics

Now, collect all of the components together for your Photopopper. They should look something like Figure 16-5.

Once you have gathered all of your components, you need to take your printed circuit board (PCB), and the pair of 2N3906 transistors, and solder them to the board so that the curved side of the transistor matches the curved symbol on the circuit board. You can see that the transistors go either side of the area marked "Trim Pot."

This is illustrated in Figure 16-6.

Now we move on to Figure 16-7, where we will solder the trimmer potentiometer in between the two transistors that we have just added to the PCB. In this circuit, our trimmer potentiometer acts as a "steering wheel" for our robot—we need to calibrate it so that our robot travels in a straight line, which allows us to manually calibrate our robot. The leads will only allow the component to go in one way, so it is hard to get wrong.

We are now going to install the two diodes. Be careful when handling these components, as they are made of glass, and so are fairly fragile. You need to observe the stripe printed on the glass casing—diodes will only work correctly if they are inserted the right way round, so make sure that you get this right.

If you have bought the Solarbotics PCB, you will see that the stripe on the board denotes which way to solder this particular component (Figure 16-8).

Next take the pair of 1381 voltage triggers, and solder these in at the top of the board. They need to be installed in a similar manner to the transistors, with the curve matching the legend on the PCB. Take note of the orientation of the components in the picture to ensure that the flat sides are correctly oriented (Figure 16-9).

There are now two capacitors which need to be soldered at the very top of the circuit board. These capacitors are not electrolytic capacitors, so it

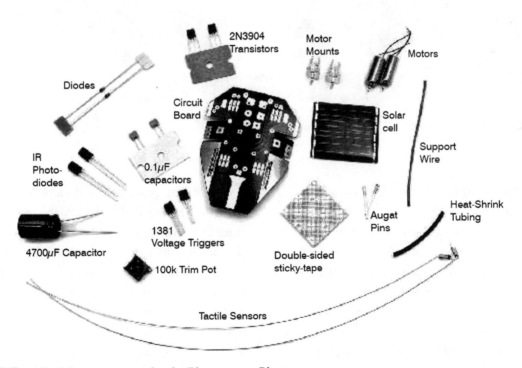

Figure 16-5 *All of the components for the Photopopper Photovore.*

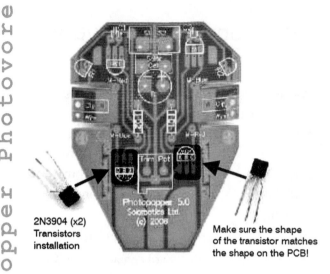

2N3904 (x2)
Transistors
installation

Make sure the shape
of the transistor matches
the shape on the PCB!

Figure 16-6 *The transistors—the first stage of assembling your Photopopper Photovore.* Image courtesy Solarbotics.

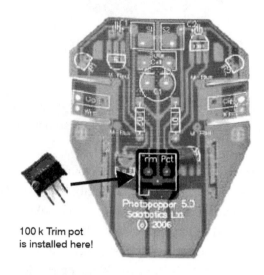

100 k Trim pot
is installed here!

Figure 16-7 *Adding the trimmer potentiometer.* Image courtesy Solarbotics.

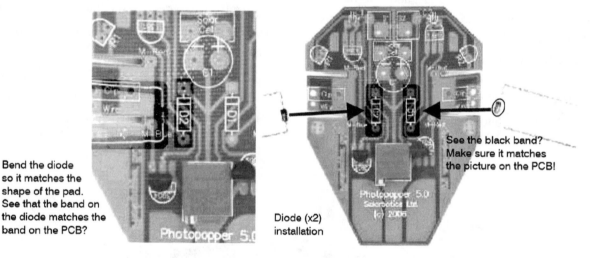

Bend the diode
so it matches the
shape of the pad.
See that the band on
the diode matches the
band on the PCB?

See the black band?
Make sure it matches
the picture on the PCB!

Diode (x2)
installation

Figure 16-8 *Installation of the diodes.* Image courtesy Solarbotics.

doesn't matter which way round they are soldered into the two holes. If you are having trouble locating the capacitors, take a look at Figure 16-10.

Now take the two optical sensors, and solder them so that the sensing element (which is the curved face) is facing outwards as shown in Figure 16-11. The optical sensors are photodiodes, which allow a variable amount of power to flow to the capacitor depending on how much light is hitting them.

Now we are going to install the 4,700 µF power capacitor (Figure 16-12). Unlike the two smaller capacitors that we installed earlier, this one is polarity sensitive as it is an electrolytic capacitor, so we need to make sure that we install it the correct way round. The capacitor will have a stripe in light blue, with minus symbols running down next to one lead. The PCB also has a pad marked with a minus symbol—so these two need to go together. You also need to make sure that the

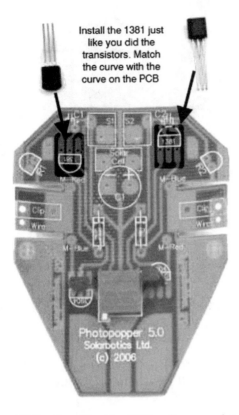

Install the 1381 just like you did the transistors. Match the curve with the curve on the PCB

Figure 16-9 *Installing the 1381 voltage regulators.* Image courtesy Solarbotics.

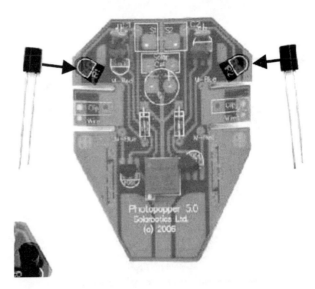

Figure 16-11 *Installing the photodiodes.* Image courtesy Solarbotics.

Figure 16-12 *Installing the power capacitor.* Image courtesy Solarbotics.

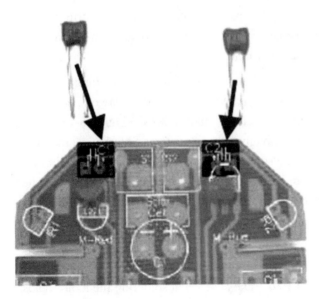

Figure 16-10 *Inserting the two capacitors.* Image courtesy Solarbotics.

capacitor lies flat against the circuit board; so you will need to bend the leads at 90°, once you have ascertained which way round the capacitor is going.

The next step is to attach the motor mounts for the tiny motors which will provide the movement for our robot. This is shown in Figure 16-13. These motor holders are in fact fuse clips. You need to solder them to the two small tabs which stick out of the side of the flexible circuit board. There is a slight complication here, in that rather than soldering the other side of the circuit board where the pin protrudes, you need to solder on the same

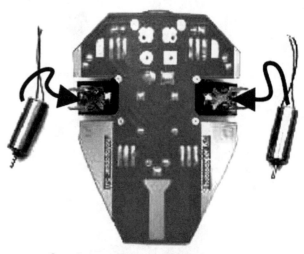

Install the motor mounts <u>on top</u> of the circuit board at these locations. Make sure the <u>tabs</u> are towards the inside before soldering or bending over the motor mount leads!

Figure 16-13 *Soldering the motor clips onto the PCB.* Image courtesy Solarbotics.

Snap the rear of the motor into the motor clip and snug it up to the tab

Figure 16-15 *Putting the motors in place.* Image courtesy Solarbotics.

Figure 16-14 *Both motor mounts ready.* Image courtesy Solarbotics.

side as the clip (Figure 16-14). This can be a little tricky.

Some fuse clips come with a little lip molded into the metal spring in order to try and retain the fuse. If the fuse clip you buy is of this type, you need to be able to solder the clip onto the board with this side facing the center line of the printed circuit board. If you don't do this, then it will be next to impossible to seat the motor comfortably as

the little lip will prevent the motor from locating properly.

Now, you will notice that the printed circuit board is a little bent—this is how it is supposed to be; however, in order to keep it rigidly bent, a piece of supporting wire needs to be soldered to support and maintain tension. You want to take the stiff piece of wire, strip some insulation off one end, and solder in the place where there is a hole marked "wire" near the motor clip. Now, pull the piece of wire to the other motor clip, strip the insulation from that end, and solder it in place to the other hole.

Once the motor clips are in place (and you have given them a while to cool down!) you need to think about the small motors (Figure 16-15). If you have bought the motors from Solarbotics, they will come supplied with a red wire and a blue wire. It is important that the correct polarity is observed when soldering the motors, as failure to do so will result in incorrect operation of your Photopopper.

The PCB from Solarbotics clearly labels the attachment point on the circuit board M-Blue and M-Red. Take care when soldering the motor leads, as they are delicate and the motor insulation is easily damaged by the soldering iron.

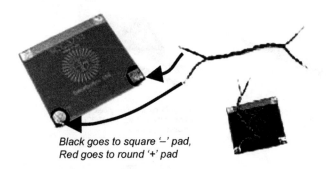

Black goes to square '–' pad,
Red goes to round '+' pad

Figure 16-16 *Connecting the solar cell.*
Image courtesy Solarbotics.

Also, if you need to strip any insulation from the wires, do so very carefully as it is very easy to break the connection between the motor and wire—and a lot harder to fix it back again!

Next comes connection of the robot's powerplant— the solar cell. First of all, take your soldering iron and tin the pads of the solar cell. Then solder a short length of wire to the pads. At this point, in order to provide a little strain relief for the wires and the carefully soldered joints, you might want to take a little hot melt glue or epoxy, and glue the wires to an area of the solar cell (for example the area between the solder pads) where there is no solder.

You will notice that one of the pads of the solar cell is round, and the other is square. The round pad is positive and the square pad is negative. So solder the black to the square and the red to the round. This is shown in Figure 16-16.

Then, take the wires and, ensuring polarity, solder them to the PCB (Figure 16-17).

The same convention as used on the solar cell has been used with the pads on the PCB—the round one is a positive terminal and the square one a negative.

This should be a real Frankenstein's monster moment, as your robot will now start to twitch with the first indications of light! The circuit is now complete, and as the capacitor charges the motors will start to whirr!

Solder the black wire to the Photopopper square pad,
and the red wire to the Photopopper round pad

Figure 16-17 *Connecting the solar cell to the PCB.*
Image courtesy Solarbotics.

Peel off one backing, and
fold the DSST in half

Peel off the other half, then squish it
hard onto the back of the solar cell!

Figure 16-18 *Attaching the solar cell.*
Image courtesy Solarbotics.

Now what you need to do is stick the solar cell to the PCB using a double-sided sticky pad (Figure 16-18).

Now take a short 10 mm length of heatshrink tubing and slip it over the tiny shaft of one of the motors. Using a source of heat, such as a lighter or match, shrink the tubing onto the shaft.

Now comes the fiddly fine-tuning—there is a small trimmer potentiometer (trim pot) that you

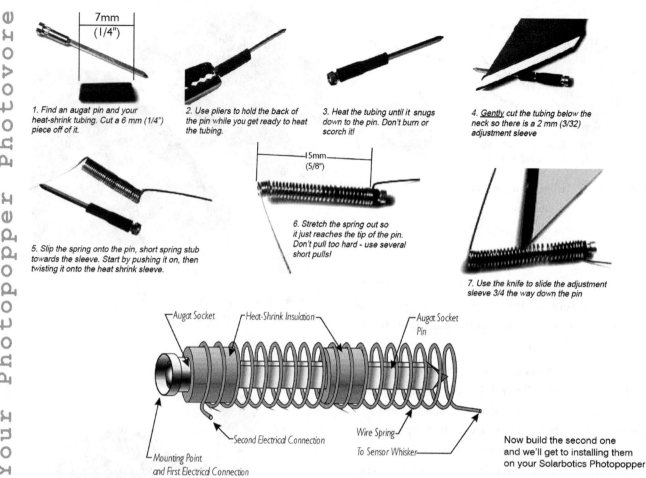

1. Find an augat pin and your heat-shrink tubing. Cut a 6 mm (1/4") piece off of it.

2. Use pliers to hold the back of the pin while you get ready to heat the tubing.

3. Heat the tubing until it snugs down to the pin. Don't burn or scorch it!

4. _Gently_ cut the tubing below the neck so there is a 2 mm (3/32) adjustment sleeve

5. Slip the spring onto the pin, short spring stub towards the sleeve. Start by pushing it on, then twisting it onto the heat shrink sleeve.

6. Stretch the spring out so it just reaches the tip of the pin. Don't pull too hard - use several short pulls!

7. Use the knife to slide the adjustment sleeve 3/4 the way down the pin

Augat Socket

Heat-Shrink Insulation

Augat Socket Pin

Second Electrical Connection

Wire Spring

To Sensor Whisker

Mounting Point and First Electrical Connection

Now build the second one and we'll get to installing them on your Solarbotics Photopopper

Figure 16-19 *Assembling the spring/socket assembly.* Image courtesy Solarbotics.

installed in the center. Well think of this as a bit of a manual steering wheel for our Photopopper.

What this trim pot essentially does is adjust the amount that our robot veers to the left or to the right. So we can use it to compensate for any stray variables that could cause our robot to track off center.

Take a small jeweller's screwdriver to adjust the trim pot. Turn the screw at least 20 times to the left—you should find that only the left motor activates. Now take the screwdriver and turn it the other way. You will find that the other motor activates. Now turn the trimmer 10 turns to center it—both motors should turn equal amounts at this point.

If your robot persistently wanders to one side or the other, then turn the screw in the direction of the engine which needs more power.

Now we are going to assemble the touch sensors for our robot. Take the small Augat sockets, and a 7 mm length of heatshrink tubing, and shrink the tubing onto the socket.

Now trim the tubing with a sharp knife below the neck of the socket and slide off the excess. Taking the spring, stretch it a little, and push the Augat connector into the center of the spring. Follow Figure 16-19 which illustrates the stages of constructing the spring/socket assembly.

Once you have completed this stage, you will want to think about soldering it to the circuit board. You need to solder the pin to face forward as shown in Figure 16-20, and then the spring wire to the adjacent pad.

Now you can curl the wires in different configurations. Experiment with making the wires different

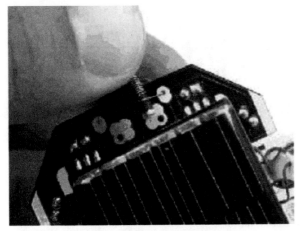

Arrange sensor so the pin head is on the big pad, and the sensor spring stub goes to the small stub

Finished sensor spring installation

Figure 16-20 *Attaching the touch sensor to the PCB.* Image courtesy Solarbotics.

shapes and seeing how it affects their ability to sense. Moving the wires to different positions will allow the robot to sense different zones in front of it. Experiment to see which arrangement makes the robot navigate most effectively.

Building your own solar robots

Armed with these simple principles you can go on to create your own solar robots.

The key to the design is the solar motor schematic and interconnection of lots of small simple "neuron" circuits.

Figure 16-21 *Solar motors are high-efficiency.*

Try and buy high-efficiency motors (Figure 16-21), as they will produce the best performance from relatively modest solar cells.

You might find old scrap electrical devices to be a great source of small motors! Look around for

Taking it further

If you want to find out more about BEAM robotics, I can highly recommend the publication *Junkbots, Bugbots & Bots on Wheels* by Dave Hrynkiw and Mark Tilden, published by McGraw-Hill.

battery-powered, or low-voltage home devices which contain motors sufficiently small to be driven by solar power. Devices containing tape decks yield a variety of small motors (Figure 16-22), so look out for old Walkmans, Dictaphones, and answering machines especially.

Figure 16-22   *Electrical devices are a great source of motors!*

Chapter 17

Solar Hydrogen Partnership

"Fuel cell technology is so appealing that it will have an enormous impact across all energy markets." K. Atakan Ozbek, Allied Business Intelligence Senior Analyst.

One of the problems that we have seen with renewable energy is "intermittency." Unlike coal, gas, and oil, where the amount of energy we produce is determined by the amount of fossil fuel we input into the process, things with renewable energy are slightly different—we must live with the energy that the weather gives us.

This presents us with a choice. We can tailor our energy demands to when the energy is available. In some situations this is practical; however, in most situations, there is an urgency to our demand for energy.

We can meet this demand in a number of ways. We can connect together renewable energy sources that are distributed over a wide area, "a National Grid" as it were. While this is helpful, it is not a complete solution—we hope that there will be sun in one area when there is not in another, but we cannot guarantee it! Furthermore, by transmitting electricity over long distances, there is a certain amount of "loss" inherent in the system. The electricity gets "lost" as a result of resistance, decreasing the total amount of useful energy available.

So, what do we do? Storing the energy would seem like an obvious candidate. But the problem is that present battery technology is fairly heavy, inefficient, and expensive.

In steps . . . the Fuel Cell.

The fuel cell—and the hydrogen economy it entails—is put forward by many as the answer to our impending energy crisis.

The hydrogen economy entails a transition from carbon-rich fuels (such as the fossil fuels we use today) to hydrogen, and carriers of hydrogen such as methanol.

Electricity is generated in the usual way from renewables—a nice mixed portfolio of wind, solar, wave, and tidal energy, producing power as and when the weather permits.

So what is so great about the hydrogen economy?

The hydrogen economy could have a number of benefits, both economic and environmental. For a start, when you burn hydrogen, you don't produce carbon dioxide, which is produced with carbon-based fuels and is a major contributor to the greenhouse effect. In addition to this, you don't get any of the other nasties such as oxides of sulfur and nitrogen that you get with burning conventional fuels.

Then there are the economic benefits. At the moment, the United States, and many other countries, cannot produce enough oil to meet their demand for energy. This places those countries in a situation where they are dependent upon the Middle East and other oil-rich nations to provide the energy to power their economies. This is not a good position for a country to be in.

It has bad consequences, as it means that in order to get oil, a desperate country might resort to any means within its power to secure that vital supply of fossil fuel. This could even include war.

By contrast, there is nothing about hydrogen that says it must be produced in a particular location.

As long as the two fundamental factors are present—electricity and water—it can be produced anywhere.

One of the great advantages of the "hydrogen economy" is that it works very well with decentralization. To understand what decentralization and decentralized power generation means, think "don't put all your eggs in one basket." To explain this a little further, at present, power is generated in bulk in large centrally managed power stations. This has been so because up until now (we have been kidding ourselves) there has been an abundance of energy-dense fuels that we can use to meet our power needs.

Now envisage a world where rather than polluted atmospheres and mountains of toxic nuclear waste, we instead have "distributed" energy production. This means solar cells on roofs, and wind turbines here and there, integrating with the fabric of our towns and cities. All of these devices produce a small amount of power—but the key is *they produce it where it is needed.*

How will fuel cells penetrate our lives in the future?

You can expect to see fuel cells in all the places you see rechargeable batteries at the moment—and many more. Fuel cells offer the prospect of laptop computers that can be used many times longer than conventional batteries allow, mobile phones with much longer periods of talk time than is possible with present battery technology, and vehicles with clean emissions of nothing but water.

"I believe fuel cell vehicles will finally end the hundred-year reign of the internal combustion engine as the dominant source of power for personal transportation. It's going to be a winning situation all the way around—consumers will get an

efficient power source, communities will get zero emissions, and automakers will get another major business opportunity—a growth opportunity." William C. Ford, Jr, Ford Chairman, International Auto Show, January 2000.

"Fuel cell vehicles will probably overtake gasoline-powered cars in the next 20 to 30 years." Takeo Fukui, Managing Director, Research and Development, Honda Motor Co., *Bloomberg News*, June 5, 1999.

So is there only one type of fuel cell?

No, there are lots! Each type of fuel cell is suited to different applications. Broadly speaking, fuel cells can be divided into two main kinds—high temperature and low temperature.

Figure 17-1 shows the main areas of focus in fuel cell technology, where research capital is being invested.

We are going to be experimenting with a type of low-temperature fuel cell—the PEM fuel cell which stands for Polymer Electrolyte Membrane, or Proton Exchange Membrane depending on who you listen to!

What is a fuel cell made up of?

If we were to take a fuel cell to bits, we would see that it is mechanically very simple. Take a peek at Figure 17-2.

We can see that the fuel cell has two end plates. These are used as a casing for the fuel cell. They help contain the internal elements; furthermore, they provide an interface to the connection for hydrogen and oxygen gas.

Next up are the electrodes. These are the pieces that allow us to "tap off" electricity. They are

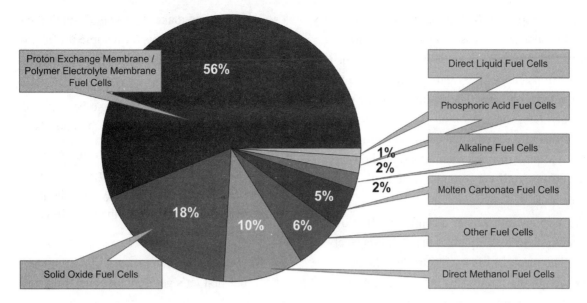

Figure 17-1   *Fuel cell technology focus.*

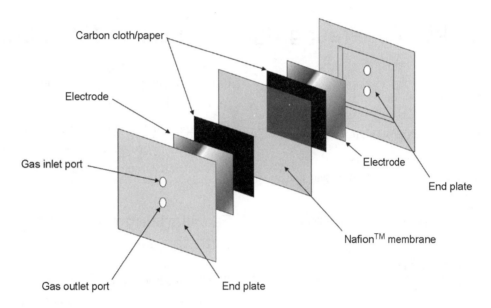

Figure 17-2   *Fuel cell construction.*

generally made from stainless steel, as it does not react with the chemicals present. The stainless steel is perforated to allow the gas to penetrate.

The next assembly of carbon cloth/paper and Nafion membrane is termed an MEA, or membrane electrode assembly. This is the bit that makes the reaction take place that produces the energy.

Next we have a carbon cloth or paper. The gas can permeate this quite easily, one side—the side

that interfaces with the Nafion membrane (more on that later)—contains a quantity of platinum which acts as a catalyst for the reaction that will take place next.

Now the Nafion membrane—but first a little explanation!

For those of you who were wondering, Nafion is a sulfonated tetrafluoroethylene copolymer. Are you any the wiser? Well Nafion is a special kind of plastic developed by DuPont in the 1960s. It's very special properties are essential to the operation of the PEM fuel cell. Essentially, what happens is that electrons cannot pass through the Nafion membrane but protons can. The platinum on the carbon cloth facilitates the separation of the electrons from the protons in the hydrogen atom. The protons are allowed to pass through the membrane; however, the electrons can't get through. Instead, they take the next easiest route.

The carbon cloth acts as a conductor. It allows the electrons to find a path to the stainless steel mesh. The mesh forms the electrode, which is connected to the circuit that the fuel cell powers. The circuit presents the route of least resistance, so the electrons make their way through the circuit. As they do this they perform some useful work which we can harness.

On the other side of the Nafion membrane is a mirror image assembly, with another set of carbon cloth, stainless steel electrode, and end plate.

At the other side, the electrons are reunited with the protons that have passed through the membrane, and the all-essential oxygen. The protons, electrons, and oxygen combine to form H_2O—better known as water.

We will look at this process in more detail later, but first let's get on with generating some hydrogen!

Project 45: Generating Hydrogen Using Solar Energy

You will need

- PEM reversible fuel cell (Fuel Cell Store part no. 632000)
- Photovoltaic solar cell (Fuel Cell Store part no. 621500)
- Gas storage tank (2 ×) (Fuel Cell Store part no. 560207)
- Rubber tubing
- Distilled water (not just purified!)

Tools

- Crocodile clip leads
- Syringe

Fuel cell tech spec

PEM reversible fuel cell

$2 \times 2 \times \frac{1}{2}$ in. (5 × 5 × 12.5 cm)

2.4 oz. (68 grams)

0.95 volts open circuit

350 MA

In this project, we are going to look at the potential for a solar-hydrogen economy.

We are going to start with a simple experiment to generate hydrogen using a solar cell to provide the electricity to electrolyze water.

Familiarizing ourselves with the stuff!

If you have bought the items above, the chances are that you have got a lot of cool stuff, but are none too sure what to do with it. Don't panic!

We are going to look at what the stuff does, and how it all goes together in this section.

First is our fuel cell, shown in Figures 17-3 to 17-5.

First of all, you should note two terminals on the top—one red, one black. It should be apparent that these are the supply terminals for the fuel cell. Then, if we look on either side of the fuel cell, we see that there are intake pipes for gas. There should be two, these are diagonally offset in the fuel cell specified above.

Figure 17-3   *PEM fuel cell with caps.*

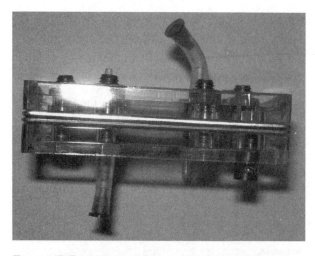

Figure 17-5   *PEM fuel cell with rubber hoses.*

Figure 17-4   *PEM fuel cell without caps.*

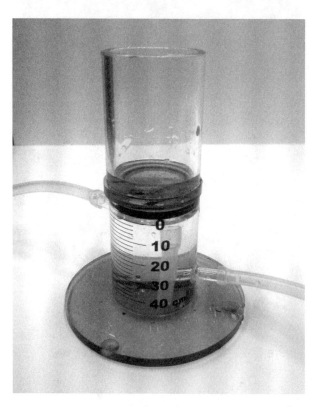

Figure 17-6   *40 ml "gas tank" filled with water.*

You will see that one side of the fuel cell has a label "H2," this is the hydrogen side; the other side has the label "O2," this is where the oxygen goes.

The fuel cell comes supplied with some little caps as shown in Figure 17-3. These can be used to prevent water from escaping. These caps can be removed if desired, and two little plastic tubes are exposed for connection to the gas tanks (more about this later). Small lengths of rubber

tube are shown attached to the fuel cell in Figure 17-5.

Next we have the "gas tanks" shown in Figures 17-6 and 17-7.

We can see how in the first instance we fill the gas tanks with water—this is illustrated in Figure 17-6. Then, as shown in Figure 17-7, as our fuel cells produce gas, the gas displaces the water which goes into the top half of the cylinder. The weight of this body of water acting on the gas provides a little pressure on the gas—enough to speed its return to the fuel cell.

We also connect the other pipe to the tank to enable the excess water to return.

The fuel cell is connected mechanically as in Figure 17-8. Note that the oxygen feed is connected in the same manner as the hydrogen.

Figure 17-7   *40 ml "gas tank" filled with gas.*

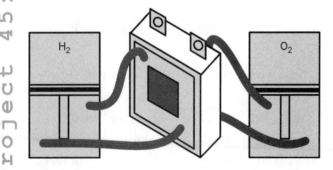

Figure 17-8   *How to connect the pipes to the fuel cell.*

Preparing the fuel cell for electrolysis

Before we can electrolyze water, we need to prime the fuel cell with water so that it has something to electrolyze. For this, we will be using distilled water. It is important that the water that you use is "distilled water," which should be readily available from the drug store, not just "purified" water. Water from the tap, even water from the chiller, contains trace elements that have the potential to wreak havoc with the delicate little MEA inside our PEM fuel cell.

First of all, you need to prime the fuel cell with water. A syringe and some small bore rubber tubing helps you accomplish this easily. Fill the fuel cell through one of the holes, allowing air to escape from the other. Once you have done this, put the caps back on the gas intake tubes of the fuel cell to prevent any ingress of air.

Fill the gas cylinders with water as well, and then connect it all up as shown in Figure 17-8. If there are any little gas bubbles trapped in the pipes, these must be bled out of the system first of all.

Now we come to connecting our solar cell.

Wiring up the solar cell and fuel cell

Once the "mechanical engineering" is complete, we need to work on the "electrical engineering." Luckily, connection is very simple indeed. Take a peek at Figure 17-9 which shows how it is done.

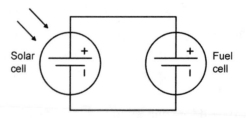

Figure 17-9   *Connecting the two cells.*

Logic should tell you that the red terminal on the fuel cell is positive and the black terminal is negative. Use your crocodile leads to connect things up and put your solar cell somewhere where it will receive good light.

Observation

Don't hold your breath, the process is going to take a little bit of time! Over time you should observe a number of things happening. Gas will begin to form, and is collected in the two cylinders, displacing the water as it does so. You will notice that twice as much hydrogen is produced as is oxygen.

What is happening here?

The chemical symbol for water is H_2O. Those with even a smattering of knowledge of chemistry will know that this means a water molecule is comprised of two hydrogen atoms and one oxygen atom.

When we pass an electrical current through the water, using the reversible fuel cell as our electrolyzer, we are splitting the water into its constituent parts—hydrogen and oxygen.

Because of the membrane inside the fuel cell, the hydrogen and oxygen are kept separated. The hydrogen and oxygen can then be piped off, and stored in tanks, where the hydrogen and oxygen can be saved for later use.

Other methods of generating hydrogen

There are also a number of other proposed methods for generating hydrogen from solar energy, which, although we are not going to examine them in detail here, are certainly of merit.

In fact, at the moment, most hydrogen is produced using a process known as steam reforming. This takes fossil fuels and combines them with steam, which disassociates the hydrogen from the carbon, However, do not be fooled! This process still leaves a whole lot of carbon dioxide to be gotten rid of.

At the moment, there is talk of a technology called "sequestration." Sequestration involves separating the carbon dioxide emissions from industry or an item of plant. They are then piped to a facility where they can be forced underground into natural geological features which supposedly will keep the carbon locked up under the earth's surface. Put simply, this technique involves hiding the problem by burying it underground. It is a good technology in so far as it allows the oil companies to maintain the status quo by continually producing oil. However, the carbon dioxide is still present—albeit "sequestered" underground. There is also another incentive for companies to adopt sequestration—it increases their oil production! By forcing gas underground, more oil comes to the surface. Economically it makes good sense, environmentally the case is yet to be proven. No one has yet tried sequestration on the scale proposed before—it is unknown whether this gas will find ways of escaping back to the atmosphere.

One method proposed for making hydrogen using solar energy, is to use bacteria to produce it. Using a process of photosynthesis, certain algae and bacteria have been shown to produce hydrogen gas. This hydrogen can be harnessed and used to power the hydrogen economy.

Online resources

This is a good website to check out that talks about the bioproduction of hydrogen.

www.energycooperation.org/bioproductionH2.htm

Project 46: Using Stored Hydrogen to Create Electricity

You will need

- PEM reversible fuel cell (Fuel Cell Store part no. 632000)

- Photovoltaic solar cell (Fuel Cell Store part no. 621500)

- Gas storage tank (2×) (Fuel Cell Store part no. 560207)

- Rubber tubing

- Distilled water (not just purified!)

- Small load (only a volt or two)/variable resistor

Tools

- Ammeter

- Voltmeter

We have seen in the previous experiment, how we can use electricity, more specifically "solar power," to generate hydrogen. We also saw that this hydrogen can be stored for later use. Now, in this part of the experiment, we are going to look at turning this hydrogen into electricity, and what is actually happening.

In the hydrogen economy, hydrogen which is generated from surplus, cheap renewable energy, can be distributed through a network of pipes and used to provide power in the home via a small residential generator (which produces useful heat as a byproduct). Additionally, this hydrogen can be used to run motor cars or public transportation systems if stored in tanks on the vehicle.

We are now going to see how a fuel cell converts this hydrogen back into electrical power.

Figure 17-10 gives a schematic representation of our fuel cell, showing the major parts.

If we look at Figure 17-11 we can see hydrogen entering the fuel cell—ready to do its thing!

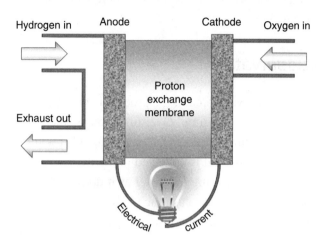

Figure 17-10 *Schematic representation of a PEM fuel cell.*

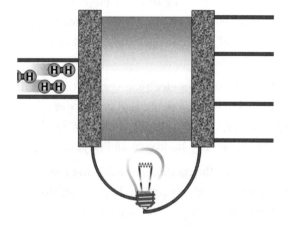

Figure 17-11 *Hydrogen enters the fuel cell.*

A reaction occurs as a result of the platinum catalyst. The electrons cannot go anywhere, as they cannot penetrate the membrane formed by the MEA, so they go round a circuit because this is the easiest path for them to take. Meanwhile, the protons escape across the membrane. This is illustrated in Figure 17-12.

As the electrons travel around the circuit, they do some useful work. For example, in Figure 17-13 they are lighting a bulb.

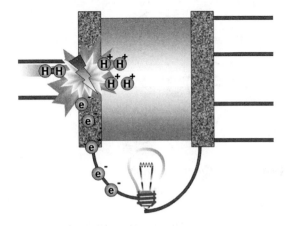

Figure 17-12 *The protons and electrons separate.*

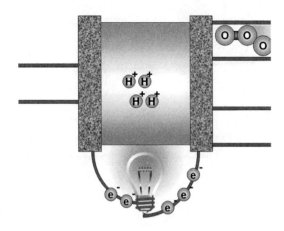

Figure 17-13 *The electrons "do some work."*

At the other side of the membrane, the electrons (which have passed around the circuit), the protons (which have crossed through the membrane), and oxygen (from either the air or oxygen tanks), are brought together, where they react. You can see this in Figure 17-14.

The product of this reaction is water, H_2O (Figure 17-15).

This process is happening continuously rather than in discrete stages. All the time there is a never ending ceaseless flow of electrons, protons, hydrogen, and oxygen.

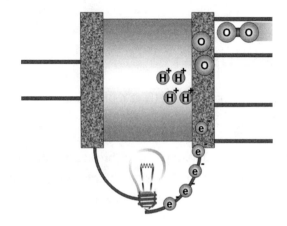

Figure 17-14 *The electrons, protons, and oxygen are reunited.*

Connecting up your fuel cell

We now need to connect up our fuel cell as shown in Figure 17-16. The mechanical setup remains the same as for the last project. However, rather than being connected to a solar cell, we are connecting our fuel cell to a load, so that energy can be extracted.

We are monitoring this load with a voltmeter and ammeter.

Instantly, you should see the amount of hydrogen and oxygen in the tanks decrease little bit by little bit and bubbles flow through the pipes. Your load, if it is a small bulb, motor, or just a resistor, should show some sign of activity. The voltmeter and ammeter will confirm that you are producing power.

Well done—you have made another great step toward the world understanding the hydrogen economy and applying it in practice!

Note

For a demonstration in an educational setting, the Eco H_2/O_2 system available from Fuel Cell Store (see Supplier's Index) part no. 534407 offers a great way to demonstrate the principles in this chapter, in a nice, desk-mounted study of solar hydrogen electricity.

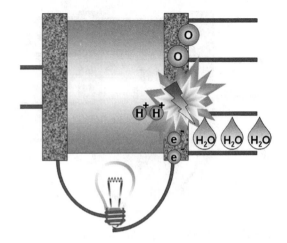

Figure 17-15   *Water is produced.*

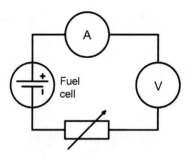

Figure 17-16   *Circuit diagram.*

Conclusion

I like to think of fuel cells as a bit like a sandwich. If you can, imagine a cheese sandwich, with the two pieces of bread as the electrodes. Those pieces of bread are buttered. It is this butter that effects the interaction between the cheese and the bread—the butter can be likened to the gas diffusion media.

Online resources

This website gives a particularly good animation that clearly illustrates what is happening inside a PEM fuel cell.

www.humboldt.edu/~serc/animation.html

The hydrogen economy isn't something that is going to happen overnight. Furthermore, it won't be a single event where all of a sudden we switch from one to the other. Instead, hydrogen technologies will gradually begin to permeate our lives.

This will probably start first of all with vehicles and portable electronic devices, as these devices will benefit from lightweight, high-density energy storage that hydrogen affords. We can expect to see hydrogen in more and more places. However, there are technological hurdles that must be overcome first of all.

Photosynthesis—Fuel from the Sun

We can harness solar energy in a lot of different direct ways, as we have seen already in this book. We can use the sun to meet our requirements for heat and electricity. Sometimes, as we saw in Chapter 17, we need energy sources that are portable and lightweight—for powering cars, for example, and for transporting energy to sites where it is not appropriate to use solar energy directly.

One way in which we can harness solar energy, is to produce plants that we can turn into fuels. Remember, plants use the sun's energy to grow. A plant takes in carbon dioxide from the atmosphere, water, and nutrients from the soil, and turns it into biological matter. The trees and flowers would not exist without the sun's energy. We see in Figure 18-1 the cycle that takes place.

Some plants can be turned into oils. Remember, each time you eat some French fries, they were cooked in oil which came from vegetable matter, often sunflower. This oil can be burned directly in engines.

It is also possible to turn vegetable oils into a product called "biodiesel." Biodiesel can be used in most ordinary diesel engines as if it were any other diesel fuel. The difference between the vegetable oil, which is a triglyceride, and the biodiesel, is that the biodiesel contains shorter hydrocarbon chains. This means, that among other properties, the oil is less viscous, which means that it flows more freely.

Another way that we can produce biofuels is to produce sugar crops, which can be fermented and distilled to produce ethanol. This ethanol can be

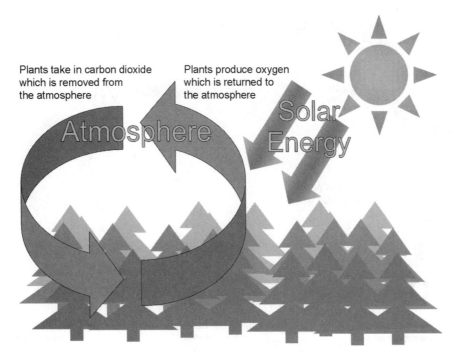

Figure 18-1 *Biofuel—fuel from the sun.*

used in gasoline/petrol engines with relatively little modification.

Countries like Brazil often use a blend of ethanol derived from plant matter, and gasoline/petrol from conventional sources. This is sold as "Gasohol" in many places. By blending the gasoline/petrol with ethanol, they are reducing their country's dependence on imported oil. The U.S.A. and many countries of the "developed world" can learn a lot from these developing countries, as the developed world is now largely dependent on imported fuels from the Middle East. The oil that is available on "home turf" is now largely located in areas of outstanding natural beauty, places that are very sensitive to environmental change. U.S. oil companies are now having to cut pipelines through swathes of Alaska, causing enormous environmental damage. Surely there are other solutions—solutions that can benefit U.S. farmers?

If we look at how ethanol (Figure 18-2) is produced, in Figure 18-3 we can see that the process is driven by solar energy. However, it is important to note, that at other stages there are inputs of energy. These might not always come from renewable energy sources, and so it is important to be critical and consider how much "oil" might be in your biofuel.

Some farming techniques rely on intensive use of fertilizers and other agricultural chemicals—all

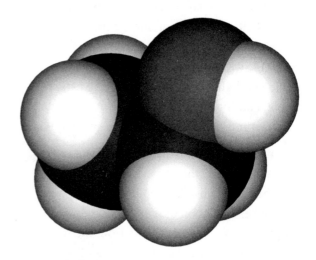

Figure 18-2   *A molecule of ethanol—a biofuel from the sun.*

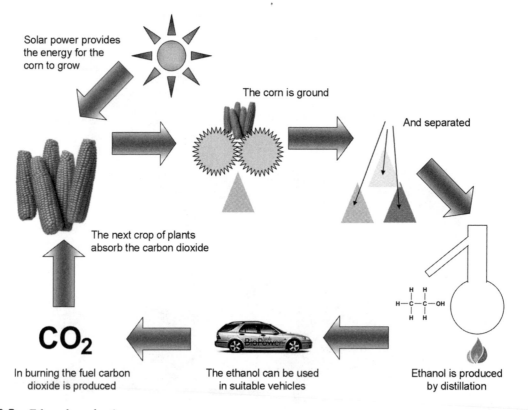

Figure 18-3   *Ethanol production.*

of these require a vast input of energy to manufacture. However, there are other farming techniques, such as organic farming, which rely on natural, low embodied energy techniques. The techniques have been honed and refined over hundreds of years, and passed down through the ages (remember industrialized farming is only a relatively recent phenomenon).

According to C.J. Cleveland of Boston University, annual photosynthesis by vegetation in the U.S.A., is 4.7×10^{16} Btu, equivalent to nearly 60% of the nation's annual fossil-fuel use. This suggests that the amount of plant matter growing is of a similar order of magnitude to the amount of energy that we consume. It would not be feasible to meet all of our energy needs from biofuels, but it certainly suggests that they could be used in a wider range of applications.

Hydrocarbons

Hydrocarbons are chains of hydrogen atoms and carbon atoms.

Coal comes from periods in the world's history when there were lots of trees and plants growing, which died, and then were compressed by rocks and earth. The carboniferous matter in the coal (which means "containing carbon"), is actually dead plants. Thus, we can see that coal was also produced from solar energy.

The crucial difference to recognize between coal, fossil fuels, and the fuel that we call biomass, is that the biomass fuel has recently taken the carbon dioxide, which is produced when it burns, out of the atmosphere. By contrast, fossil fuels, when burnt, are releasing the carbon dioxide which was stored under the ground for many millions of years. That is why, when we burn fossil fuels, we really are creating a massive problem.

The four hydrogen rule

Let's not get too hung up about chemistry here, but at the same time, let's try and understand how hydrocarbons work. By now, if you have been paying attention, you should realize that atoms of carbon and hydrogen combine to form molecules, which we call hydrocarbons. This much was explained in the "Hydrocarbons" box. However, let's try and understand the rules for the combination of hydrogen and carbon atoms.

A carbon atom has four "sites" where other atoms can potentially bond. In the simplest hydrocarbon, methane, all four of these sites are bonded to hydrogen atoms. We might also know this chemical by the name "natural gas." However, in addition to hydrogen atoms, carbon atoms can also bond to these sites.

We can make chains of carbon atoms by removing one hydrogen bond from each of two methane molecules, and joining the carbons together by this "missing" bond.

We can make these chains longer and longer, until eventually we start getting to the point where we have eight carbons in a row. When we have eight carbons joined by single bonds, we call this "octane" and for this level of chemistry, we can understand that "octane" is similar to petrol or gasoline. If we start adding carbons, we get to the point where we have between 10 and 15 in a row. The diesel we use in our cars is a mixture of 10–15 carbon long hydrocarbon chains.

Our ecosystem is a bit like a bucket with a hole in it under a tap. The tap is like the carbon dioxide we are putting into the atmosphere, and the hole in the bucket is like the mechanisms which remove carbon dioxide from the air—the plants and vegetation which absorb CO_2 and release oxygen.

If we pour water into the bucket at the same rate that it flows out, then the level in the bucket stays constant.

If we are burning mainly biofuels, and replanting the things that we burn, then we are only putting carbon dioxide into the atmosphere which was removed recently. This is like pouring water in at the rate it flows out.

However, if we start turning the tap so that the water gushes out, then the level in the bucket begins to rise.

At present, that is what we are doing—we are turning up the tap on the carbon dioxide in our atmosphere. We are letting it gush out of the tap rapidly, with the effect that it is now beginning to get to the point where we have more water in the bucket than it can hold safely!

This is because we are putting carbon into the atmosphere which was safely tucked away millions of years ago.

If you look at Figure 18-4, you can see that the plant is taking in water and carbon dioxide and producing oxygen. However, those with some knowledge of chemistry must realize that any equation must be balanced. Take a peek at Equation 1.

$$6H_2O + 6CO_2 \Rightarrow C_6H_{12}O_6 + 6O_2$$

Equation 1 *The photosynthesis equation.*

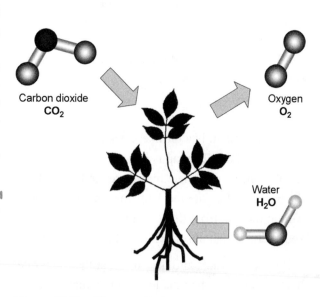

Carbon dioxide
CO_2

Oxygen
O_2

Water
H_2O

Figure 18-4 *Photosynthesis.*

Here we can see that water and carbon dioxide combine to produce glucose and oxygen. The glucose is the "food" for the plant that enables it to grow.

A snapshot history of biofuel

Since the earliest days, when an intrepid caveman rubbed some sticks together and discovered he could warm himself, and finally cook that hot meal he had been dying for, we have been burning biofuels. Wood has been a staple fuel for heating and cooking.

With the advent of the steam age (the invention of the "external combustion engine"), it became possible to turn heat into kinetic energy—motion. This paved the way for the industrial revolution, and brought us mechanized transport in the form of early steam vehicles.

Much of the industrial revolution was powered by coal, as it was easy to extract and had a high energy density. However, there are also numerous examples of steam engines being powered by wood fuel.

This is all very nice . . . but things really start to get interesting with the invention of the internal combustion engine. This invention can be attributed to one Nikolaus August Otto (June 14, 1832–January 28, 1891) whose picture you can see in Figure 18-5. His idea was pretty revolutionary, and really threw the cat amongst the pigeons. Rather than burning the fuel "outside" of the cylinder, the new idea was to burn the fuel "inside" the cylinder. In May 1867, there was a new revolution, the internal combustion engine was born.

You might think that this is bad news for biofuels, as it could be tricky squeezing large logs inside a small cylinder. Quite the contrary in fact, Otto's original plan was to use ethanol, which we have read about in this chapter. Ethanol is a biofuel.

Otto's company is still ticking over nicely and now exists under the name Deutz AG.

In a four-stroke, Otto cycle, internal combustion engine, a spark ignites the mixture of fuel and air. These are the type of engines you find in petrol or gasoline cars.

However, there is more to the story than that! Biofuels got another boost when one Henry Ford designed his mass-production "Model T" car to run on ethanol! Unfortunately now, our story takes a bit of a sinister turn.

During the Second World War, with supplies of oil scarce, countries began to look at using biofuels to meet the war-effort's insatiable demand for energy.

Unfortunately, after this time biofuels take a bit of a turn for the worse. Oil became cheap and biofuels disappeared into obscurity . . . until now.

The western world is now finding it harder than ever to find fossil fuels at acceptable prices—both financial and environmental. As a result of the world's insatiable lust for oil, areas of outstanding natural beauty such as Alaska are being plundered for their oil, with dire environmental

Figure 18-5 *Nikolaus August Otto.*

consequences. There is a resurgence of interest in biofuel technology—expect to hear *much* more about biofuels in the years to come.

Bad biofuels?

Well, all Evil Geniuses should be trying to form a balanced view of the arguments in the world they are trying to conquer. Let's take a look at the flip side of biofuels and assess why, maybe, they aren't going to save the world.

Biofuels have an important role to play while we look for medium-term solutions to our energy problems. In the longer term, technologies such as the hydrogen economy and hydrogen fuel cells, could potentially meet our energy needs. However, in the short term, we need to look for transitional solutions that will allow us to shift from our "dirty" technologies, to more environmentally responsible technologies. Biofuels have a part to play in this transition.

However, if we look at trying to meet all of our energy needs from biofuels, it becomes apparent that "growing energy" may not be desirable on a large scale. There is limited bio-productive land (by which we mean land with the ability to produce crops) in the world. We need some of this land to live on, we need some of this land to produce food to eat, and we need some of this land for animals to graze. If you do the math, and look at a future energy scenario where our needs are met by biofuels, you find that it just doesn't add up.

With rising populations, and hence rising demand for food, and furthermore, rising expectations from developing countries, it is apparent that at the moment there isn't enough land to meet our needs *wholly*.

This doesn't matter in the short-term, as there is plenty of land that is yet to be put to good use, and plenty of "biologically derived" waste products of industry, that have the potential to provide us with energy if we see them as a product rather than something to be disposed of.

Furthermore, we need to look at products that are currently treated as waste that could be used to provide energy. There are fast-food outlets on every block that produce French fries by the million. In doing so, they consume copious amounts of vegetable oil, which eventually ends up as spent oil to be disposed of. Did you know that you can run a diesel engine on waste vegetable oil with minimal effort?

> Watch the book sellers shelves, my forthcoming book *Convert Your Vehicle to Biodiesel in a Weekend* will contain details of how you can convert a diesel vehicle to run on vegetable oils.

However, while running your vehicle on a waste product might be saving the world in your own little way, cutting down rainforests to plant fuel crops is not. Shamefully, this is what is happening in some parts of the world, which sort of defeats the object of biodiesel being a more sustainable fuel.

Vast areas of rainforest are being devastated to plant palm oil crops that produce quick oil—and money—for the growers. However, for every acre of rainforest we lose, we also lose a massive amount of biodiversity—and a little bit of the earth's lungs.

Moreover, planting large amounts of the same thing, isn't such good news for biodiversity. We need a healthy mix of different flora and fauna—having large overwhelming amounts of the same thing isn't particularly good for the ecology of the world.

Photosynthesis experiments

In the following experiments, we will be analyzing the solar-driven processes that happen in plants, to produce usable products which we can burn as biofuels. The process which we have explored already, is called "photosynthesis."

Sugarcane, rapeseed, and other crops which are generally used to produce biofuel are a bit big, hard to manage, and unpredictable. For this reason, we will be modeling the production of biofuel

crops using a small, more manageable plant: salad cress/mustard. To give you an idea of what you should be looking for at the garden center, here are the seeds that I used, pictured in Figure 18-6.

Note

In order to conduct accurate experiments, we need to control all of the variables as closely as we can, in order to make sure that our experiments are accurate and repeatable. Although it might seem a bit pedantic, try and ensure as far as possible that you give the plants the amount of water, cotton wool, light, etc. as stated in the text. To ensure a fair test, try and make sure that all things remain the same. For example, if you put several tests on a windowsill, ensure that they all receive equal light, and one is not in shadow and one in light.

Figure 18-6   *Cress seeds suitable for our experimentation.*

Project 47: Proving Biofuel Requires Solar Energy

You will need

- 40 salad cress seeds
- Cotton wool (cotton batting)

Tools

- Two small bowls
- 5 ml measuring syringe

In this experiment, we are going to test the hypothesis that "biofuel requires solar energy in order to be produced." In order to do this, we are going to set up two small test cells, which we will use to compare the growth of two samples of our "biofuel," cress.

Take two small bowls. Fill the bottom of one with a layer of 2 cm of cotton wool. Carefully pick out 20 cress seeds from the packet, and water them evenly with 5 ml (5 cm^3) of water. We want to make our two cells as identical as possible; be as accurate as you can, as in order for the test to be fair, it must be repeatable.

One of these bowls should be placed on a bright sunny windowsill. The other should be placed on the same windowsill but covered with a dark box, to prevent any sunlight from reaching the seeds.

Observe the seeds over several days to see what happens. Compare the two samples, but remember to cover the "dark" seeds soon after taking a peek.

Your results should confirm the hypothesis that solar energy is required for seeds to grow.

Of course, we can mimic the properties of sunlight using artificial sources of light, but this would seem self-defeating as it requires a lot of energy.

Project 48: Proving Biofuel Requires Water

You will need

- 40 salad cress seeds
- Cotton wool

Tools

- Two small bowls
- 5 ml measuring syringe

This experiment is very similar to the previous one, in that we will be comparing two different containers of "biofuel crop." The difference with this experiment is that both will be exposed to bright sunlight. However, one will receive 5 ml (5 cm^3) of water, while the other will be on dry cotton wool.

The results of this experiment should seem intuitive—if you don't water plants, they die; however, the experiment is worth carrying out nonetheless.

Project 49: Looking at the Light-Absorption Properties of Chlorophyll

You will need

- Four identical boxes
- 80 salad cress seeds
- Cotton wool
- Colored filter gelatin red, green, and blue

Tools

- Four small identical bowls
- 5 ml measuring syringe

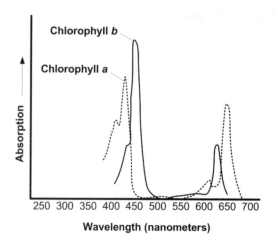

Figure 18-7 *Chlorophyll's response to light.*

Note

If you are not sure where to get filters from, ask in a good photographic suppliers or stage lighting shop for colored gelatin. If this does not work, you could always use the clear colored plastic that some sweets are wrapped in, carefully joined with sticky tape—ensuring that the joins are good and no additional light penetrates.

We have seen in a previous experiment that for the photosynthesis process to occur, light is an essential component. We are now going to look at this light in a little more detail.

We are going to set up our cress with the light and water that they require. However, this time, we are going to have a "control" experiment, and three other boxes where the sunlight they receive is filtered to "red," "green," and "blue."

Take four boxes. Fill the bottom of each one with a layer of 2 cm of cottonwool. Carefully pick out 20 cress seeds from the packet for each box, and water them evenly with 5 ml (5 cm³) of water.

You will need to attach a colored gelatin filter to the top of three of the boxes, so that the only light which reaches the cress seeds is colored.

Now, compare the growth of the cress seeds. What seeds seem to be doing well? Why do you think that is?

Chlorophyll in plants is a "photoreceptor." It is what converts the light from the sun to food for plants to grow. In green plants, there are "chloroplasts" which contain chlorophyll—these are green in color and account for the green coloring of plants.

There are two types of chlorophyll, a and b; they both respond to similar wavelengths of light, as can be seen in Figure 18-7.

You will probably find that the red and blue filtered plants grew well, whereas the green filtered plants did not. If you look at Figure 18-8, you will see that the wavelengths where the "peaks" are correspond to the wavelengths of blue and red colored light.

The chloroplasts absorb red and blue light, but reflect green light—that is why we see plants

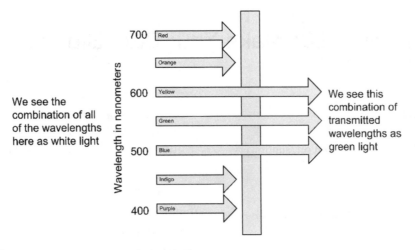

Figure 18-8　*Green light as a component of white light.*

as green. During the fall/autumn season, the photosynthesis activity decreases and the level of chlorophyll drops—that is why we see the leaves turn from green to reds and orange!

Figure 18-9　*Rudolph Diesel.*

Factoid

Remember that we said plants take in carbon dioxide (CO_2), and produce oxygen (O_2)? Well, let's put some statistics to that assertion! One hectare, which is just under two and a half acres, of corn, produces enough oxygen to meet the requirements of around 325 people!

Biodiesel

At the moment, our cars run on gasoline and diesel from deep oil wells. Because we are fast running out of oil, we see the prices steadily creep up. Well, what if oil grew on trees. Well, it might not grow on trees, but biodiesel certainly comes close!

Like many of the ideas in this book, it is not a new one. Rudolph Diesel, inventor of the diesel engine, pictured in Figure 18-9, designed the original compression ignition engine (that's diesel engine to you and me) to run on a wide range of hydrocarbon fuels. In 1898, Diesel demonstrated his engine running on peanut oil!

We can run compression ignition engines on a wide variety of vegetable oils directly, with a little modification. However, because these oils are quite thick (viscous), problems can be encountered if the fuel is cold. For this reason, we take ordinary fuel, and turn it into biofuel with a little chemical ingenuity.

As plants produce their food using photosynthesis, biodiesel can be thought of, in a way, as "liquid sunshine" solar energy stored in the chemical bonds of plants, ready to be used at will.

Project 50: Make Your Own Biodiesel

You will need

- 100 ml vegetable cooking oil (corn oil, sunflower oil, etc.)
- 20 ml methanol
- 1 g of lye (sodium/potassium hydroxide)

Tools

- Glass lab flask, or cylinder
- Glass rod to mix
- Hydrometer
- Safety equipment
- Eye wash
- Goggles
- Gloves
- Apron
- Vinegar

Warning

Because of the methanol and the lye used in this experiment (both of which are toxic and not nice to handle) this experiment should be carried out under the supervision of a responsible adult. Lye is a *really* strong alkali, which can cause chemical burns if you are not careful. If you get the dust in your eye, it can cause permanent damage, even blindness. You may think that it is an unusual inclusion for a safety list, but in the event of spilling any lye, put a spot of vinegar on the lye to neutralize it and make it safe. The acid will react with the alkali and produce carbon dioxide. Make sure you wear all the safety equipment above and take sensible precautions.

If you are wondering where to get lye from, it is commonly sold as caustic drain cleaner, but be aware, it is nasty stuff so take safety precautions.

If you want to know where to get methanol from, try a good modeller's shop—as it is often sold as a fuel for model plane engines.

In this experiment, we are going to be making biodiesel, a fuel that will run in diesel engines, produced from biological matter, in this case vegetable oil. It is possible to run an ordinary diesel vehicle on biodiesel as can be seen in Figure 18-10. Some environmentally responsible companies are beginning to run fleets of vehicles on biodiesel. Increasingly, we are seeing blends of biodiesel and petro-diesel being sold on garage forecourts, and in

Figure 18-10 *Filling a car with biodiesel.*

some cases, small quantities of biodiesel are being added to ordinary diesel as a lubricant.

In this project, we are going to make a small quantity of biodiesel. This experiment aims to show the process, and illustrate a little bit of the chemistry of manufacturing biodiesel. However, please note that there is very little "quality control" in this experiment, so it would be unwise to use the produced biodiesel in a diesel engine.

First take the vegetable oil and heat it up to over 100°C to ensure that any water is driven off. Be careful not to overheat.

In a separate container, thoroughly mix the methanol and lye. This forms a mixture that biodiesel home brewers call "methoxide," you need to be *very* careful during this stage as it entails using some very nasty chemicals.

Now add the methoxide to the vegetable oil. Try and do this at around 45°C for the best reaction.

Be careful not to breathe in any of the methanol vapors. If you have access to a fume cupboard in a school chemistry department, then use that for safety. If not, just make sure you are in a well-ventilated area—outside for example.

Make sure that you thoroughly stir the mixture for a few minutes.

Now, put the container somewhere safe, out of the reach of small children, and allow the mixture to settle overnight.

When you next look at the mixture, you should see that it has settled into two distinct layers. At the bottom of the container will be a brown goo. This is a mixture of glycerol, unused methanol, and catalyst, and possibly a little soap, which will be the product of any fatty acids in the oil to begin with.

The layer on top is our biodiesel! Again, I reiterate the earlier warning—*do not* put this in your vehicle. It will probably work, but this particular biodiesel has been made with no quality control, so there is a high possibility that it could damage your engine.

Online resources

If you want to make biodiesel, you might be interested in the following web resources, which contain information on manufacturing biodiesel.
www.schnews.org.uk/diyguide/howtomake biodiesel.htm
www.veggiepower.org.uk/
journeytoforever.org/biodiesel_make.html

Testing your biodiesel

The first test that you will need to conduct is a visual inspection—what does your biodiesel look like?

Well, there should be a clear visual distinction between the top layer and the bottom layer in your measuring container. If there is a large intermediate layer of soaps, then the chances are that you did not remove sufficient water from the oil to begin with.

You may be able to get a piece of equipment called a "hydrometer." A hydrometer will tell you the density of a fluid. Your biodiesel should check out with a density of between 880 and 900 grams per liter.

Biodiesel chemistry

So what have we done in this little experiment? If you look at Figures 18-11 and 18-12, you will see that we have what is called a "triglyceride" molecule. This is the vegetable oil that we start off with. We can represent this using a "ball and stick" model, which shows us the position of the atoms in 3D space, or we can show the chemistry using a "structural formula" such as that shown in Figure 18-12.

You will notice when looking at the structure of the triglyceride that there are three long "chains" of hydrocarbons (hence the 'tri' bit) and a "bacbone,"

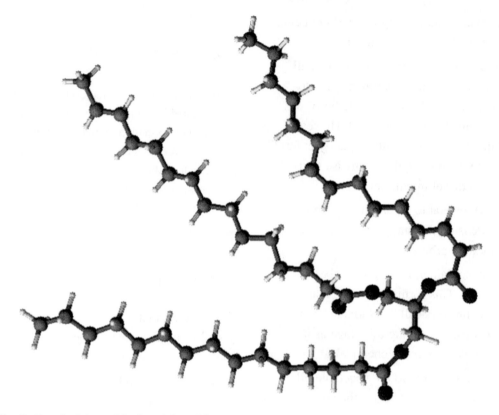

Figure 18-11 *Ball and stick model of a triglyceride.*

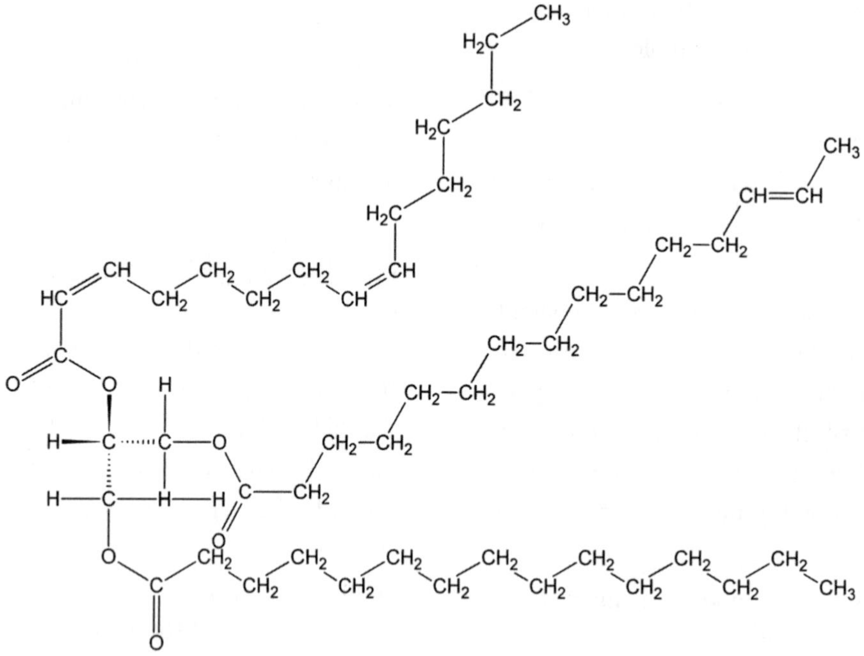

Figure 18-12 *Structural formula of a triglyceride.*

which links up the three hydrocarbon chains. This backbone breaks off from the chains to form "glycerol," and this accounts for the "glyceride" bit of the name.

Our methoxide mixture which was added to our triglycerides catalyzed a reaction where the three chains broke off from the backbone. The glycerol backbone is heavy, so settles to the bottom of the jar, whereas the lighter hydrocarbon chains float to the top of the jar—and it is these which form our "biodiesel."

Why do we separate the chains from the glycerol backbone? Well, the structure of the triglyceride means that it can easily become "tangled up" with other triglyceride molecules, which results in a thick viscous oil. The chains on their own, are a lot shorter and so tangle less easily, and as a result, are less viscous. This brings with it a whole host of other desirable properties, which means that the fuel can flow freely in a diesel engine without clogging up the injectors.

Appendix A

Solar Projects on the Web

Hopefully, this book has provided an interesting primer on solar energy. I rather hope that I have empowered you with the tools to go onwards and create your own solar inventions, powered by clean green energy. To help you on your way, and provide inspiration, I have compiled a list of projects, articles, and inspirational solar tech stories that I have rooted out for you on the web. This should provide creative inspiration for your own sustainably powered designs. All of these sites seem to be particularly interesting or quirky—this highlights the need to be creative when finding applications for technologies, rather than being constrained by the dogma of existing applications. I truly hope they will inspire you, the reader, to go out there and create something truly unique.

Want Java? Go Solar

I've seen it all now—a solar-powered coffee roaster! Clean energy when used with fair-trade coffee makes for a guilt-free expresso.
www.makezine.com/blog/archive/
solar_roast_coffee.jpg
www.solarroast.com/home.html

Solar-Powered Wiggly Sign

In this fantastic site, engineer and cartoonist Tim Hunkin builds an interesting piece of kinetic sculpture, powered by the sun.
www.timhunkin.com/a125_arch-windpower.htm

Humungous Solar-Powered Laundry

The world's biggest laundry employs solar power! Read all about how clothes are being washed without the negative environmental connotations at *USA Today.*
www.usatoday.com/tech/news/techinnovations/
2006-07-30-solar-laundromat_x.htm?csp=34

Solar-Powered Wheelchair

Bob Triming introduces us to his clean green form of transport—a solar-powered wheelchair. Flat batteries are a thing of the past.
www.infolink.com.au/articles/63/0C044763.aspx

Villages in Tanzania Use Solar Power to Sterilize Water

Here is an example of the solar still technology seen in this book in action. Villagers are using it to purify their water and prevent illness in Africa.
news.bbc.co.uk/1/hi/world/africa/4786216.stm

Solar Clothing Monitors Medical Conditions

Certainly not the coolest looking jacket on the block, this solar-powered item of clothing monitors medical conditions by using inbuilt sensors. Hopefully, you won't feel ill in the dark.
www.digitalworldtokyo.com/2006/09/
taiwan_puts_ehealth_solar_pane.php

Solar-Powered iPod Shuffle

Powered by a flexible solar cell, *Make:* show how to convert an iPod shuffle to run on green power.
www.makezine.com/blog/archive/2005/03/
solar_powered_i.html

Massive Parabolic Solar Death Ray

This death ray was built for cooking hot dogs at the Burning Man event. The scale of this thing is totally immense! It is well worth a look.
igargoyle.com/archives/2006/07/solar_death_ray_
for_hot_dogs.html

MIT Solar Generator from Car Junk

In trying to invent a low-cost form of solar energy, these academics at the Massachusetts Institute of

Technology showcase their solar generator from old auto junk. Recycled green power!
www.technologyreview.com/read_article.aspx?id=17169&ch=biztech

Solar Water Warmer
A simple and inexpensive design for a solar water heater. Brought to you by *Mother Earth News*.
http://www.motherearthnews.com/library/1979_September_October/A_Homemade_Solar_Water_Heater

Sun-Powered Tunes
Listen to some music while not having the guilt of carbon emissions! A solar-powered boom box—because let's face it, who wants a party on an overcast day?
www.makezine.com/blog/archive/2006/05/homemade_solar_powered_boombox.html

Solar Sunflower
This 'bot, which is essentially a solar tracker, follows the sun in the same way as sunflowers do. Don't water it though!
www.instructables.com/id/E8UMC79GJAEP286WF5/

Organic LEDs Generate Power
Organic LEDs—a term you are going to hear about a lot more in the future! As well as producing light from electricity, these clever little gadgets also have the potential to produce power from light.
www.ecogeek.org/content/view/242/

Solar Scooter Provides Sustainable Transport
A home-built solar-powered scooter, which during a last look at its odometer had clocked up near to 1,000 miles. Scoot on!
www.treehugger.com/files/2005/09/diy_eco-tech_ti.php

Solar Networking
Interesting article about solar-powered wireless networking in Boulder, Colorado! Is this the future of the Internet?
www.internetnews.com/wireless/article.php/3525941

Solar Death Ray
Another solar death ray, with a cool gallery of burnt stuff! Remember kids, if you toast your siblings, your parents may get mad at you. In addition, note the obligatory: "Do not stare into the beam directly."
www.solardeathray.com/

Solar Bike Light
Bicycles are already a nice sustainable method of transport that do not chuck plumes of carbon dioxide into the air. Well, go one step further, rather than powering your lights from batteries that go in the bin, build a solar-powered bike light. Genius!
www.creekcats.com/pnprice/Bike05-Pages/bikelight.html

Solar Greenhouse
A nice page about the joys of adding a solar greenhouse to the side of a dwelling.
www.theworkshop.ca/energy/grnhouse/grnhouse.htm

Solar-Powered Handbag
One for the girls, a solar-powered handbag, designed by a smart student from Brunel University, U.K. The solar panel charges a battery, which powers a lining, which glows when the bag is opened. Never lose your keys again! Louis Vuitton nothing!
news.bbc.co.uk/2/hi/technology/4268644.stm

Passive Solar Collector
A nice passive solar collector made from wholly recycled materials.
www.theworkshop.ca/energy/collector/collector.htm

Solar Projects on the Web

Solar Ant-Zapper

Possibly cruel (that's my disclaimer to avoid letters from the American Humane Association) but this solar-powered ant-zapper could be the answer to your infestation woes.
www.americaninventorspot.com/
backyard_solar_energy

Organic Solar Cells Based on Biological Mechanisms

We read in Chapter 18 all about the process of photosynthesis, but how about organic solar cells which mimic the processes that plants use to produce energy. Interesting!
www.abc.net.au/science/news/stories/s1729572.htm

Honda Prius Supplements Power with Solar

A modified Honda Prius, which uses solar power to supplement the juice in the car's batteries to bring an increase in economy!
www.treehugger.com/files/2005/08/
solar-powered_t.php

Solar-Powered Light Graffiti

For miscreants who like to tag, here is the perfect substitute, which does no damage, and uses clean renewable energy to make its mark.
rdn.cwz.net/archives/17

Holographic Solar Cells

Read about holographic solar cells, which use a holographic optical element to concentrate the frequencies that matter!
www.prismsolar.com/

Top 10 Strangest Solar Gadgets

This blog provides some interesting thought-provoking ideas—the top 10 strangest solar gadgets. Will you invent number 11?
www.techeblog.com/index.php/tech-gadget/
top-10-strangest-solar-gadgets

Solar Ferry for Hyde Park

A solar-powered ferry for the Serpentine in Hyde Park, London, U.K. How novel! Forty-eight foot long with 27 panels on its roof.
www.usatoday.com/tech/science/
2006-07-18-solar-ferry_x.htm?csp=34

Solar-Powered Backpack

Charge your laptop with the power of the sun while trekking up Everest! Now we have heard it all
www.rewarestore.com/product/020010003.html

Sun Bricks

An interesting concept, solar-powered bricks that incorporate light emitting diodes to provide illumination at night—a novel idea!
www.gardeners.com/on/demandware.store/Sites-
Gardeners-Site/default/ViewProductDetail-
SellPage?OfferID=35-945&SC=xnet8102

Proposed Solar Chimney Down Under

This massive solar chimney has been proposed to generate power in Australia, using a large greenhouse to heat air, and a solar thermally driven process. All of 1,600 ft tall!
money.cnn.com/2006/08/01/technology/
towerofpower0802.biz2/index.htm?cnn=yes
www.enviromission.com.au/project/technology.htm

Nanosolar—Printed Solar Cells

If Google founders Larry Page and Sergey Brin stick their cash here, you can bet that it is big business. Nanosolar are a new venture who aim to produce lots of cheap solar cells using simple printing technology.
nanosolar.com/index.html

Supplier's Index

Miscellaneous

Alternative Energy Hobby Store
Dennis Baker
49732 Chilliwack Central RD
Chilliwack BC, V2P 6H3
Canada
Tel: 1-604-819-6353
Fax: 1-604-794-7680
Dennis@AltEnergyHobbyStore.com
www.altenergyhobbystore.com/
Education_solar_books.htm

Arizona Solar Center
c/o Janus II—Environmental Architects
4309 E. Marion Way
Phoenix AZ 85018
USA
solar@azsolarcenter.com
www.azsolarcenter.com/bookstore/reviews.html

Centre for Alternative Technology Mail Order
Machynlleth
Powys
SY20 9AZ
UK

Kentucky Solar Living
Tel: 1-859-200-5516
kentuckysolar@ipro.net

Silicon Solar
Direct Sales
Tel: 1-800-653-8540 (Mon–Fri 8 am–4 pm EST)
Fax: 1-866-746-5508

Tampa Bay, FL
Tel: 1-727-230-9995
Fort Worth, TX
Tel: 1-817-350-4667
San Diego, CA
Tel: 1-858-605-1727
www.siliconsolar.com/solar-books.php

Solar Electric Light Fund
1612 K Street, NW Suite 402
Washington DC 20006
USA
Tel: 1-202-234-7265 (8:30 am–6 pm EST)
info@self.org
www.self.org/books.asp

Drinking birds for solar engines

The Drinking Bird
Tel: 1-800-296-5408
www.thedrinkingbird.com/

HobbyTron.com
1053 South 1675 West
Orem UT 84058
USA
Tel: 1-801-434-7664
Toll-free: 1-800-494-1778
www.hobbytron.com/.html

Niagara Square
7555 Montrose Road
Niagara Falls ON, L2H 2E9
Canada
Tel: 1-905-354-7536
Fax: 1-905-354-7536

Science eStore
5318 E 2nd Street #530
Long Beach CA 90803
USA
http://www.physlink.com/eStore
1-888-438-9867
11am -8pm EST

Electronic components for the projects in this book

Electromail/RS Components
RS Components Ltd
Birchington Road, Corby
Northants NN17 9RS
UK
Orderline: 44-(0)-8457-201201
www.rswww.com

Maplin Electronics Ltd
National Distribution Centre
Valley Road
Wombwell, Barnsley
South Yorkshire S73 0BS
UK
www.maplin.co.uk

Radio Shack
300 RadioShack Circle
MS EF-7.105
Fort Worth TX 76102
USA
Tel: 1-800-843-7422

Rapid Electronics Ltd
Severalls Lane
Colchester
Essex CO4 5JS
UK
Tel: 44-(0)-1206-751166
Fax: 44-(0)-1206-751188
sales@rapidelec.co.uk
www.rapidelectronics.co.uk/

Fresnel lenses and parabolic mirrors

Alltronics
PO Box 730
Morgan Hill CA 95038-0730
USA
Tel: 1-408-778-3868
www.alltronics.com

Anchor Optical Surplus
101 East Gloucester Pike
Barrington NJ 08007-1380
USA
Fax: 1-856-546-1965

Edmund Optics Inc.
101 East Gloucester Pike
Barrington NJ 08007-1380
USA
Tel: 1-800-363-1992
Fax: 1-856-573-6295

Science Kit & Boreal Laboratories
777 E. Park Drive
PO Box 5003
Tonawanda NY 14150
USA
Tel: 1-800-828-7777
Fax: 1-800-828-3299
www.sciencekit.com

TEP
International Manufacturing Centre
University of Warwick
Coventry CV4 7AL
UK

Inverters and power regulators

Omnion Power Engineering
2010 Energy Drive
PO Box 879

East Troy WI 53120
USA
Tel: 1-262-642-7200 or 1-262-642-7760
www.sandc.com/omnion/home.htm

Real Goods
360 Interlocken Blvd, Ste 300
Broomfield CO 80021-3440
13771 So. Highway 101
PO Box 836
Hopland CA 95449
USA
Tel: 1-800-919-2400
www.realgoods.com

Photochemical solar cell components

ICE, the Institute for Chemical Education
University of Wisconsin-Madison
Department of Chemistry
1101 University Avenue
Madison WI 53706-1396
USA
Tel: 1-608-262-3033 or 1-800-991-5534
Fax: 1-608-265-8094
ICE@chem.wisc.edu

Photovoltaic cells for buildings

Flagsol
Flachglas Solartechnik GmBH
Muhlengasse 7
D-50667 Cologne
Germany
Tel: 49-(0)-221-257-3811
Fax: 49-(0)-221-258-1117

Schüco International
Whitehall Avenue, Kingston
Milton Keynes MK10 0AL
UK
Tel: 44-(0)-1908-282111
Fax: 44-(0)-1908-282124

Photovoltaic modules

Advanced Photovoltaics Systems
PO Box 7093
Princeton
NJ 08543-7093
USA
Tel: 1-609-275-0599

BP Solar International
PO Box 191, Chertsey Road
Sunbury-on-Thames
Middlesex TW16 7XA
UK
Tel: 44-(0)-1932-779543
Fax: 44-(0)-1932-762533

Kyocera
8611 Balboa Avenue
San Diego CA 92123
USA
Tel: 1-619-576-2647

Siemens Solar Industries
PO Box 6032
Camarillo CA 93010
USA
Tel: 1-805-698-4200

Solarex Corporation
630 Solarex Court
Fredrick MD 21701
USA
Tel: 1-301-698-4200

Solec International
52 East Magnolia Boulevard
Burbank CA 91502
USA
Tel: 1-213-849-6401

Supplier's Index

Solar array structures, mounting hardware

Kee Industrial Products Inc.
100 Stradtman Street
Buffalo NY 14206
USA
Tel: 1-716-896-4949
Toll-free: 1-800-851-5181
Fax: 1-716-896-5696
info@keeklamp.com

Kee Klamp GmbH
Voltenseestrasse 22
D-60388 Frankfurt/Main
Germany
Tel: 49-(0)-6109-5012-0
Fax: 49-(0)-6109-5012-20
vertrieb@keeklamp.com

Kee Klamp Limited
1 Boulton Road
Reading
Berks RG2 0NH
UK
Tel: 44-(0)-118-931-1022
Fax: 44(0)-118-931-1146
sales@keeklamp.com

Leveleg
8606 Commerce Ave
San Diego CA 92121-2654
USA
Tel: 1-619-271-6240

Poulek Solar Ltd
Velvarska 9
CZ-16000 Prague
Czech Republic
Tel: 42-(0)-603-342-719
Fax: 42-(0)-224-312-981
www.solar-trackers.com

Science Connection
50 East Coast Road, #02-57

Singapore 428769
Tel: 65-65-68966
Fax: 65-623-44589
www.scienceconnection.com/Tech_advanced.htm

Wattsun
Array Technologies Inc.
3312 Stanford NE
Albuquerque NM 87107
USA
Tel: 1-505-881-7567
Fax: 1-505-881-7572
sales@wattsun.com
www.wattsun.com

Zomeworks
PO Box 25805
1011A Sawmill Road
Albuquerque NM 87125
USA
Tel: 1-800-279-6342 or 1-505-242-5354
Fax: 1-505-243-5187
zomework@zomeworks.com

Solar controllers/temperature monitors/instrumentation

HAWCO Ltd, Industrial Sales
The Wharf, Abbey Mill Business Park
Lower Eashing
Surrey GU7 2QN
UK
Tel: 44-(0)-870-850-3850
Fax: 44-0)-870-850-3851
sales@hawco.co.uk

Raydan Ltd.
The Sussex Innovation Centre
Science Park Square
Falmer, Brighton
Sussex BN1 9SB
UK
Tel: +44-(0)-1273-704442
Fax: +44-(0)-1273-704443
sales@raydan.com

Solar pool-heater manufacturers

Heliocol
13620 49th Street North
Clearwater FL 33762
USA
Tel: 1-727-572-6655 or 1-800-79-SOLAR
(1-800-797-6527)
http://www.heliocol.com/

Imagination Solar Limited
10–12 Picton Street
Montpelier
Bristol BS6 5QA
UK
Tel: 44-(0)-845-458-3168
Fax: 44-(0)-117-942-0164
enquiries@imaginationsolar.com

Solar Industries Solar Pool Heating Systems
1940 Rutgers University Boulevard
Lakewood NJ 08701
USA
Tel: 1-800-227-7657
Fax: 1-732-905-9899
www.solarindustries.com/

Solar Twin Ltd
2nd Floor, 50 Watergate Street
Chester CH1 2LA
UK
Tel: 44-(0)-1244- 403407
hi@solartwin.com
http://www.solartwin.com/pools.htm

Solar robotics supplies

Solarbotics Ltd
201 35th Ave NE
Calgary AB, T2E 2K5
Canada
Tel: 1-403-232-6268
N. America toll-free: 1-866-276-2687

Places to get more information

American BioEnergy Association
314 Massachusetts Avenue, NE
Suite 200
Washington DC 20002
USA
www.biomass.org

American Council for an Energy Efficient Economy
1001 Connecticut Avenue, Suite 801
Washington DC 20036
USA
www.aceee.org

American Solar Energy Society (ASES)
2400 Central Avenue, Suite G-1
Boulder CO 80301
USA
Tel: 1-303-442-3130
Fax: 1-303-443-3212
ases@ases.org
www.ases.org

California Energy Commission
1516 Ninth Street
Sacramento CA
USA
Tel: 1-958-145-512, 1-800-555-7794, or
1-916-654-4058
www.energy.ca.go

Center for Excellence in Sustainable Development
US Department of Energy, Denver Regional Office
1617 Cole Boulevard
Golden CO 80401
USA
Fax: 1-302-275-4830

Energy Efficiency and Renewable Energy Clearinghouse (EREC)
PO Box 3048
Merrifield VA 22116

USA
Tel: 1-800-DOE-EREC or 1-800-363-3732
Fax: 1-703-893-0400
Doe.erec@nciinc.com

Florida Solar Energy Center (FSEC)
1679 Clearlake Road
Cocoa FL 32922
USA
Tel: 1-407-638-1000
Fax: 1-407-638-1010
info@fsec.ucf.edu
www.fsec.ucf.edu

Home Power Magazine
PO Box 520
Ashland OR 97520
USA
Tel: 1-800-707-6585
www.homepower.com

NASA Earth Solar Data
eosweb.larc.nasa.gov/sse/

National Biodiesel Board
3337a Emerald Lane
PO Box 104898
Jefferson City MO 65110-4898
USA

National Center for Appropriate Technology
3040 Continental Drive
Butte MT 59701
USA
Tel: 1-406-494-4572

National Renewable Energy Laboratory
1617 Cole Boulevard
Golden CO 80401-3393
USA
Tel: 1-303-275-3000
webmaster@nrel.gov
www.nrel.gov

North Carolina Solar Center
Box 7401
North Carolina State University
Raleigh NC 27695-7401
USA
Tel: 1-800-33-NCSUN
Fax: 1-919-515-5778
ncsun@ncsu.edu
www.ncsc.ncsu.edu

Northeast Sustainable Energy Association
50 Miles Street
Greenfield MA 01301
USA

Rocky Mountain Institute
1739 Snowmass Creek Road
Snowmass CO 81654-9199
USA

Sandia National Laboratory—California
PO Box 969
Livermore CA 94551
USA
Tel: 1-925-294-2447

Sandia National Laboratory—New Mexico
PO Box 5800
Albuquerque NM 87185
USA
Tel: 1-505-844-8066
webmaster@sandia.gov
www.sandia.gov

Solar Electric Light Fund
1612K Street NW, Suite 402
Washington DC 20006
USA

Solar Energy Industries Association
1111 N. 19th Street
Suite 260
Arlington VA 22209
USA
Tel: 1-703-248-0702
Fax: 1-703-248-0714
info@seia.org
www.seia.org/Default.htm

Solar Energy International
PO Box 715
Carbondale CO 81623
USA
Tel: 1-970-963-8855
Fax: 1-970-963-8866
sei@solarenergy.org

Index

*Page numbers for tables and figures are given in **bold***

Index

Free Catalogue and 10% off Your First Order

Request a copy of *Buy Green By Mail* and receive 10% off your first order. Send in this voucher with your order or quote CATSEPEG with your order by phone.

Voucher

Buy Green By Mail, Centre for Alternative Technology,
Machynlleth, Powys, SY20 9AZ, UK.
Orderline +44(0)1654 705959 mail.order@cat.org.uk
www.cat.org.uk/shopping

Terms and Conditions

This voucher can only be used towards payment for books and products from the Mail Order Department. It may not be redeemed for Cash. Offer ends 31st December 2009.